Pulsed Alternators Technologies and Application

Shaopeng Wu · Shumei Cui

Pulsed Alternators Technologies and Application

Springer

Shaopeng Wu
Harbin Institute of Technology
Harbin, Heilongjiang, China

Shumei Cui
Harbin Institute of Technology
Harbin, Heilongjiang, China

ISBN 978-981-33-4226-2   ISBN 978-981-33-4224-8 (eBook)
https://doi.org/10.1007/978-981-33-4224-8

This Springer imprint is published by the registered company Springer Nature Singapore Pte Ltd.
The registered company address is: 152 Beach Road, #21-01/04 Gateway East, Singapore 189721, Singapore

Preface

Pulsed power technology originated from the 1930s and received a rapid development after the 1960s. In recent years, it has been applied in many fields. High-power pulsed power supply is the main component of pulsed power system, and its growth directly affects the development and application of pulsed power technology. Pulsed alternators boast comprehensive strengths in the aspects of energy density and power density. They also integrate inertial energy storage, electromechanical energy conversion and pulse-shaping as a single element, therefore attract growing attention in the world in all fields in national economy, especially in military, as in electromagnetic emission, electromagnetic launch, microwave and laser weapons. The USA, Europe, Russia and China have launched a lot of scientific research and invested a large fortune in their researches.

During the accumulation of relative materials, the author collected related literature from top-notch research institutions in the world, especially in the USA, with a highlight on the yields and practical experience of Harbin Institute of Technology over the years in this direction. With pulsed alternators as the core component, the whole pulsed alternator system is systematically introduced, including its principle and development, its electromagnetic design, the analysis on its thermal management and mechanical performance, the control technology of pulsed alternator power system, the load and modeling of electromagnetic weapon and related technologies.

This book was written by Shaopeng Wu and Shumei Cui. In the process of writing, our appreciation is attributed to Dr. Weiduo Zhao, Dr. Xiyuan Li, Dr. Yuan Wan, Dr. Shaofei Wang, Dr. Songlin Wu, Ms Yu Luo, Mr Jinyang Zhou and Ms Xinmiao Zhang who contributed greatly in the research of pulsed power technology and pulsed alternators, and provided the precious information as we required.

In the course of writing, our gratefulness goes to the domestic and international literature we referred to, some of which have been detailed after each chapter, but some of them are still inevitably not mentioned.

Due to the limited knowledge and time, if there exist any errors or mistakes, experts and readers are more than welcome to comment and criticize.

Authors
Harbin Institute of Technology
Harbin, China

Contents

Chapter 1
Overview of High-Power Pulsed Power Supply

The research on pulsed power technology and high-power pulsed power supply was initiated in the 1930s and has developed rapidly since the 1960s. As an emerging technology, it is gradually growing into an independent discipline. High-power pulsed power techniques laid an important technical foundation for the research of national defense and high technology. It's also a cutting-edge science with promised prospect in the world.

1.1 Pulsed Power Technology

Pulsed power technology is the electrophysical technology that slowly inputs the energy of less power into the stored-energy element over a long period of time, compresses and converts the energy, and then releases it to the load at a very high power density in a very short period (can be as short as nanoseconds). It's essentially a process in which the input power is amplified by the output power.

As we know, energy is directly proportional to the product of power and time, i.e. $E = Pt$. It can be seen that when the energy is constant, after a long period of energy storage, the elements within (i.e. inductors, capacitors, etc.) will release the energy in a very short period of time with rather high amplification, as shown in the schematic diagram of pulsed power compression in Fig. 1.1.

As pulsed power technology is featured with high voltage, high current, high power, and strong pulse, the relative studies mainly focus on energy storage and the generation and application of high-power pulse, including: (1) Energy storage technology; (2) The generation of high-power pulses; (3) Pulsed switching technology; (4) High pulsed current measurement technology.

Pulsed power devices are generally composed of the following components: primary energy, intermediate energy storage, pulsed formation systems, switching

S. Wu and S. Cui, *Pulsed Alternators Technologies and Application*,
https://doi.org/10.1007/978-981-33-4224-8_1

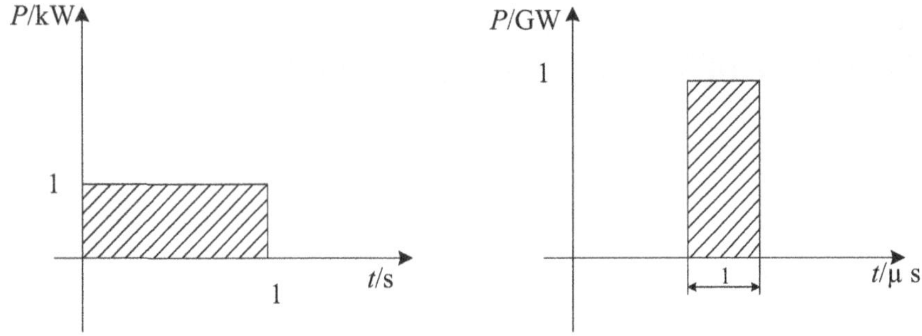

Fig. 1.1 The principle of pulsed power compression (when energy E is constant)

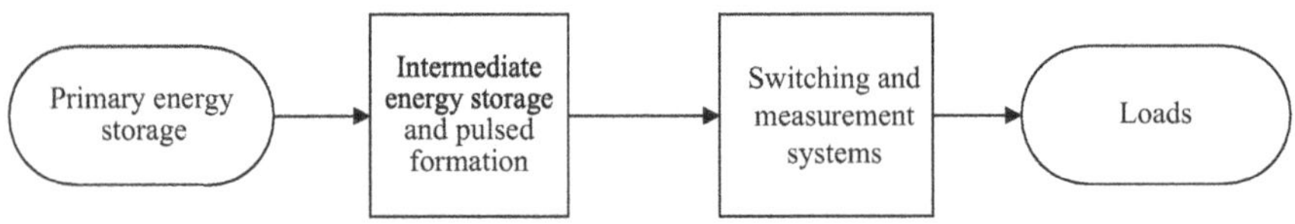

Fig. 1.2 The basic structure of pulsed alternators

and measurement systems, and loads, as shown in following Fig. 1.2. However, sometimes only switching systems but intermediate energy storage and pulsed formation systems are required.

There are a variety of ways for the primary energy of the pulsed power devices to store energy. Capacitors carry out the storage in the form of electric fields, while inductors, mechanical energy generators, chemical energy devices and nuclear energy devices achieve the goal by magnetic fields. A comparison of the four modes of energy storage is shown in Table 1.1.

The switching element plays a special role in the pulsed power system because of its parameters and characteristics which impose the most direct and sensitive effect on the rising time and magnitude of the pulse. In the pulsed power system, the traditional and frequently used switches include mechanical switches, oil-immersed switches, spark gap switches, thyratron switches, vacuum tube switches, plasma switches, discharge-included explosion conductor switches, electron beam -control reflective circuit breaker, etc. These switching techniques have been widely used in the pulsed power system. In recent years the scholars also witnessed the invention of a number of high-performance switches, such as magnetic switches, semiconductor switching devices, optical switches and so on. The main parameters of several common switches are as shown in Table 1.2.

In the pulsed power system, the high-current pulse is generally characterized by high current peak, short rising time and descent time. The main pulse is barely long but changes rapidly. In this case, pulsed current measurement is one of the key

Table 1.1 Comparison of the four modes of energy storage

Energy storage mode	Energy accumulators	Stored-energy density		Stored-energy/J	Pulsed parameters			
		J/g	J/cm3		Power/W	Voltage/V	Current/A	Pulse width/s
Capacitance	Pulsed capacitors	0 ~ 1	0.1 ~ 0.5	2.5×10^7	$10^{10} \sim 10^{14}$	$10^6 \sim 10^7$	$10^7 \sim 10^8$	$10^{-8} \sim 10^{-1}$
Inductance	Inductors	2 ~ 5	0.5 ~ 5	$10^8 \sim 10^{10}$	$10^9 \sim 10^{12}$	10^5	10^7	$10^{-3} \sim 1$
	Superconductive sensors	36 ~ 50	40 ~ 100	$10^8 \sim 10^{14}$	10^{11}	3×10^5	10^5	$10^{-3} \sim 10^{-2}$
Mechanical	Pulsed alternators	1.5 ~ 10	100	$10^9 \sim 10^{11}$	$10^9 \sim 10^{10}$	$10^4 \sim 10^5$	$10^5 \sim 10^7$	$10^{-3} \sim 1$
	Homopolar generators	3 ~ 15			10^{10}	900		$10^{-2} \sim 10$
Chemical	High explosives	5000		4.2×10^7	10^{12}	$10^4 \sim 10^6$		$10^{-6} \sim 10^{-4}$

Table 1.2 Main parameters of several common switches

	Operating voltage/kV	Peak current/kA	Switch speed/level	Repetition frequency/Hz	Life/time
Oil-immersed switch	290	3	ms	200	Short
Spark gap switch	100	40	ns	125	Short
Thyratron switch	30	50	ms	10	Medium
Vacuum switch	50	100	ns	10	Medium
Plasma Switch	4250	750	ns	100	Medium
Magnetic switch	250	40	ns	1000	Long
Gate turn-off thyristor (GTO)	6.5	140	μs	300	Long
Insulated gate bipolar transistor (IGBTs)	6.5	3	μs	150	Long
Reverse switching dynistor (RSD)	3.5	250	ns	1000	Long

pulsed power technologies, Fig. 1.3 shows the following diagram of pulsed current measurement system.

Starting with the measurement of the circuit of the current, the sensor removes a proportional amount of electrical physical quantities and sends it back to the measuring instrument via appropriate transmission.

In 1938, Kingdon and Tanis from the United States initiated the idea of using high-voltage pulsed power to produce a microsecond pulse-width flash X-ray. In

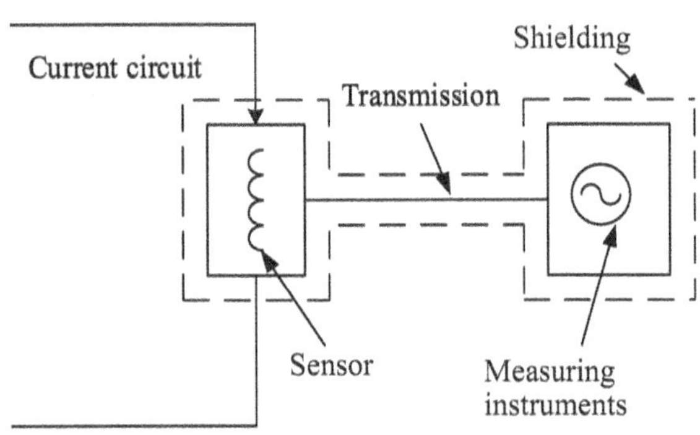

Fig. 1.3 The measurement system for high currents

1939, the Soviets managed to create vacuum pulsed X-ray tubes and applied flash X-ray photography technology to ballistics and detonation physics experiments. High-voltage pulses were obtained by charging in parallel with high-voltage pulsed capacitors and discharging in series. In 1947, A. D. Blumlien from Britain patented the law of refraction and reflection in transmission lines for pulsed forming lines, making a breakthrough in nanosecond pulsed discharge. In 1962, J. C. Martin and his team combined the Marx generator with Blumlien's patent and created the world's first intense relativistic electron beam accelerator called SMOG (3 MV, 50 kA, 30 ns) with a pulsed power of TW (10^{12} W) magnitude. Since then, the large-scale pulsed power devices advanced rapidly. In 1986 the device named PBFA-II with a peak voltage of 12 MV, a peak current of 8.4 MA, and a pulse width of 40 ns was completed. With a diode beam of 4.3 MJ and a pulse power of 10^{14} W, it was recognized as the world's first large-scale device with a pulse power exceeding 100 TW.

At present, in our country, we have world's leading institudes such as the Institute of Plasma Physics of the Chinese Academy of Sciences, the Institute of High Energy Physics of the Chinese Academy of Sciences, the Institute of Electrical Technology of the Chinese Academy of Sciences, Huazhong University of Science and Technology, and Tsinghua University. There are more than 20 Marx devices in operation in China, led by the "Flash I" built by the Southwest Institute of Engineering Physics in 1979. Later on, after the 1990s, there came "Flash II" of the Northwest Institute of Nuclear Technology, and the "Shenguang II" of the China Institute of Engineering Physics and the Shanghai Optical Machine Institute. Several typical technical parameters of the pulsed power devices are shown in Table 1.3.

Table 1.3 The comparison of the performance of the typical pulsed power devices

Model	Voltage/MV	Current/MA	Pulse width/ns	Power/TW
Hermes-I (USA)	10	0.1	80	1.0
Aurora (USA)	14	4 × 0.4	120	22.4
PBFA-II (USA)	12	8.4	40	100.8
ANGARA-I (Russia)	1	1	60	1
ANGARA-5 M (Russia)	2	0.5	100	1
A H rapa-5 (Russia)	2	40	90	80
ETIGO-II (Japan)	3	0.4	60	1.2
Raiden-IV (Japan)	1.4	1.4	50	1.96
SMOG (United Kingdom)	3	0.05	30	0.15
APEX (United Kingdom)	36	3	80	108
Flash I (China)	8	0.1	80	0.8
Flash II (China)	0.9	0.9	70	0.81

1.2 High-Power Pulsed Power Supply

The high-power pulsed power supply is the power supply that provides electromagnetic energy to the pulsed power devices. It constitutes the main body of the pulsed power device, as in almost all parts of the pulsed power device are included. The main feature of high-power pulsed power supply is the slow accumulation of energy at a rather low power before the instantaneous release of high power and large energy. If an electric pulsed device generates 10^6 to 10^9 J with a pulsed power of 10^6 to 10^{14} watts within 10^{-9} to 10^{-3} s, it's usually categorized to high-power pulsed power supply.

The high-power pulsed power supply is mainly composed of primary energy (for input), intermediate energy storage, conversion and release systems of energy (for output). The primary energy refers to low-power energy input devices, such as capacitive chargers, excitation sources for inductive coils, and driving motors of inertial energy storage motors. Their energy comes from the power grid. Intermediate energy storage devices include electric field energy storage (taking capacitors and Marx generators as examples), magnetic field energy storage (taking inductive coils at atmospheric temperature or superconducting inductive coils as examples), and mechanical energy storage based on various types of pulsed alternators with moment of inertia, chemical energy storage such as storage batteries, magnetic fluid generators, explosive magnetic flux compression generators, as well as nuclear primary energy, for instance, nuclear magnetic fluid generators. The conversion and release systems of energy are mainly composed of various large-capacity closing switches and circuit breakers and numerous waveform conditioning techniques.

High-power pulsed power technology is characterized by high voltage, high current, high power, strong pulse and high-quality waveforms. It is an interdisciplinary and comprehensive scientific topic involving motor and electrical appliances, high voltage engineering, current transformation technology, power electronics, precise electrical measurement, automatic control, relay protection, grounding technology and electromagnetism. It is endowed with significant scientific and applicable values in diversified researches about national defense and high-tech fields including nuclear explosion simulations, controlled nuclear fusion tests, high-current particle beam accelerator, high-power pulsed laser, high-power microwave, directional energy weapons, electromagnetic emission, electromagnetic propulsion, electromagnetic forming, surface treatments of materials, and ion implantation for semiconductor. At the same time, new fields with promising prospects in environmental protection such as treatments of sewage, waste gas and various harmful substances are also hot topics of recent studies.

The core technical problem of high-power pulsed power supply is pulsed- power energy storage system with high energy storage density (kJ/kg) and high- power density (kW/kg). It requires good controllability and small internal resistance of the pulsed discharge waveform to fulfill the needs of different loads. It's also a necessity to have good pulsed reproducibility and a simple system configuration. Therefore, an enhancement needs to be achieved in the aspect of density of energy storage, the frequency of repetition, the degree of lightweight, miniaturization and practicality in the future.

Western countries are investing heavily in their researches about improving the energy storage capacities of many kinds of energy storage systems. In terms of electrochemical energy storage, the United States and Japan have some progresses in the researches on high power density battery to make up for its shortcomings of the low power density of the energy storage system. The power density of lithium vitrified batteries in the United States has reached 200 kW/kg. The Japanese one owns a better performance, but the problems in heat generation and sealing of the battery are still remain to be solved.

High-energy-storage-density capacitors are developed under two directions: to develop high-strength dielectric capacitors on the one hand and electrolytic capacitors on the other hand. At present, these two types of capacitors have reached the order of magnitude of more than 10 kJ/kg. In the next five years, the energy storage density is expected to exceed 30 kJ/kg.

Mechanical energy storage involves DC generators, synchronous alternators, homopolar alternators, high-performance disc AC motors, compensating pulsed alternators, and rotary magnetic flux compressors. Different inertial energy storage manners are selected on the basis of corresponding loads.

The high-power pulsed power supply technology serves as an emerging and interdisciplinary study. Overall, it contains energy storage technology, energy conversion and power amplification, high-power switching technology, high-voltage insulation technology, and control technology.

1.3 Typical Applications of High-Power Pulsed Power Supply

1.3.1 Industrial Applications

1.3.1.1 Applications in Environmental Engineering

1. Pulsed purification of industrial waste gas

Technology of Pulsed Corona Discharge Plasma Waste Gas Purification is also known as the chemical technology of Pulse Corona Induced Plasma Chemical Process (PPCP) at ns-level. The logic behind the concept is to use a high-voltage pulse corona with a steep-front and narrow pulse width (<ns level) to discharge therefore to obtain a non-equilibrium plasma at atmospheric temperature, that is, to generate a large number of high-energy electrons and active particles such as O and OH. In this way, the deleterious molecules of industrial gas will undergo reactions such as oxidation and degradation, which will convert pollutants into low-toxic or non-toxic

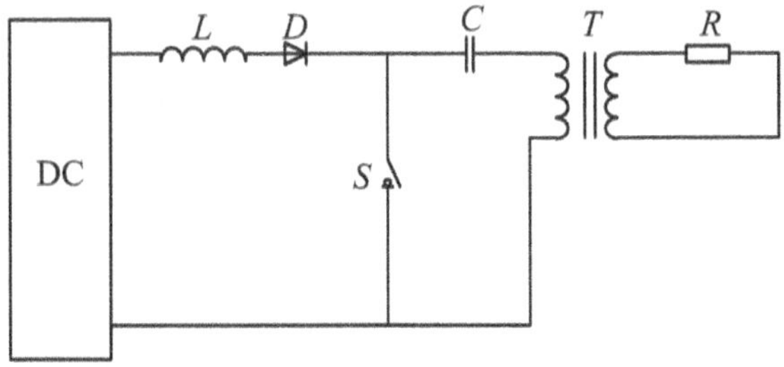

Fig. 1.4 The principle of pulsed-transformer power supply

substances. The researches of pulse corona plasma desulfurization and denitrification technology mainly focus on the development of high-voltage and narrow-pulse-width supplies, optimization of reactor structures, mechanism of chemical reactions of desulfurization and denitrification plasma, and selection of additives.

The scientific fundamental lies behind the pulsed-transformer power supply is as shown in Fig. 1.4. DC, the high-voltage power supply charges the capacitor C through the inductor L and the transformer T to the peak value, and then turns on the switch S to discharge the capacitor C therefore a pulse is formed. Pulsed-transformer current technology is more advanced, but the steepness of the pulse front is limited due to the adoption of core materials of transformers.

2. Water-treatment technology of pulsed discharge plasma

Because it is easy to achieve high energy by high-voltage technology, the water-treatment technology of pulsed discharge plasma is proposed. It is based on high-energy electron bombardment, ozone sterilization, photochemical treatment of ultra-violet light and the active free-radical generation in the discharge plasma. The research on this technology mainly focuses on the design of high-voltage pulsed power supply and the optimization of plasma generation method.

3. Pulsed electrostatic precipitation

In the process of pulsed electrostatic precipitation, the equivalent capacitance of the dust layer of the precipitator is only charged with a small amount of electric charges during the application of the pulse, and it is basically discharged during the disappearance of the pulse. There is no back corona generated from the high voltage of charge accumulation on the dust layer of the precipitator. Therefore, the precipitator of the pulsed power supply works better than the one of DC power supply in terms of the effect of dust removal. As shown in the schematic diagram in Fig. 1.5, E_p is the body of the electrostatic precipitator, and its dust absorption electrode is grounded. The isolation diode D_2 is connected to E_2, the negative electrode of the basic DC high-voltage power supply whose voltage can be adjusted. Its voltage value is usually adjusted to the point that is close to the critical corona voltage. The pulsed power supply consists of an adjustable DC voltage power supply, E_1, a filtering current-limiting inductance, L_1, a resonant energy storage capacitor, C_1, a feedback diode,

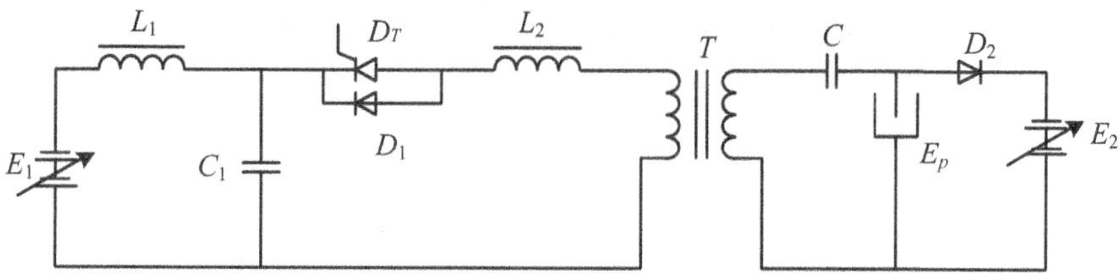

Fig. 1.5 Principle of pulsed electrostatic precipitation of pulsed power supply

D_T, of the fast switching thyristor, D_1, and a resonant circuit L_2 which composes of resonant inductors.

The pulse frequency and duty cycle can be adjusted by controlling the trigger pulse of the thyristor D_T, and the pulse width is determined by the parameters of the resonance circuit.

4. Pulsed electric field sterilization

The high-voltage pulsed electric field has an outstanding bactericidal effect on dozens of food-related microorganisms, such as E. coli, B. subtilis, Saccharomyces cerevisiae, Staphylococcus aureus, Candida faecalis, Streptococcus faecalis, and Serratia marcescens etc. The mechanism of pulsed field sterilization mainly includes cell membrane perforation effect, viscoelastic pole formation model, electromagnetic effect, electrolytic product effect, and ozone effect.

5. Pulsed preparation of ozone

The main problems of ozone technology are low production efficiency and high energy consumption. The theoretical efficiency of preparation of ozone is 2 kg/(kWh). However, the efficiency of air preparation is only 46 ~ 58 g/(kWh), and it can be raised to 10% when using pure oxygen. High-voltage pulsed technology can increase ozone production efficiency.

1.3.1.2 Applications in Oil Field

Pulsed power technology is suitable for scale removal, viscosity reduction, plugging relief and stimulation of production and injection, increased exploitation effect, and oil recovery enhancement of oil and/or water wells. For longitudinal stratus with heterogeneous properties, the ones with low-permeability and high density provides the most effective downhole discharge treatment. Therefore, this technology is mostly applicable to blocks with rapid production decline, poor reservoir utilization, low oil supply capacity, disadvantaged fracture connectivity, severe water-injection breakthrough, low water flooding efficiency, and severe blockage of crude oil emulsions. Targeting at the low porosity, oil saturation, pressure, temperature, ultra-low permeability and shallow low-productivity lithologic reservoirs in oil field in Qi Li Village, 108 wells suitable for pulsed power stimulation were selected for detonations. 2,700 more tons of crude oil were produced throughout the year. The input–output ratio was 1:2.35. The potential of the reservoir was fully tapped and a noticeable increase on yield was observed.

1.3.1.3 Applications in Electromagnetic Forming

Electromagnetic forming belongs to high-energy (high rate) forming. There are many types of high-energy (high rate) forming, but electromagnetic forming eliminates the danger of explosive forming and is more convenient than electro-hydraulic forming. From the 1950s, the technology of electromagnetic forming advanced at home and abroad and became a new processing method for plastic working of metals, which was highly valued by various industrial countries.

The principle of electromagnetic forming is shown in Fig. 1.6. Electric energy is stored in a high-voltage capacitor. When the high-voltage switch is closed, the capacitor discharges quickly into the coil (in microseconds) and provokes an abrupt change in the current in the circuit. It can be seen that a varying magnetic field will be generated around the circular coil according to the law of electromagnetic induction. With the continuous charging and discharging of capacitors, a varying pulsed magnetic field will be generated around the circular coil. When the pulsed magnetic field pierces the workpiece, an induced current (eddy current) will be generated.

Therefore, the metal workpiece becomes a charged body. According to electromagnetism, it can be known that charged metal workpieces will be affected by magnetic field forces in a drastically changing magnetic field. When such magnetic pressure equals to the yield strength of the material, the metal workpiece will undergo corresponding plastic deformation and carry out the forming of metal parts.

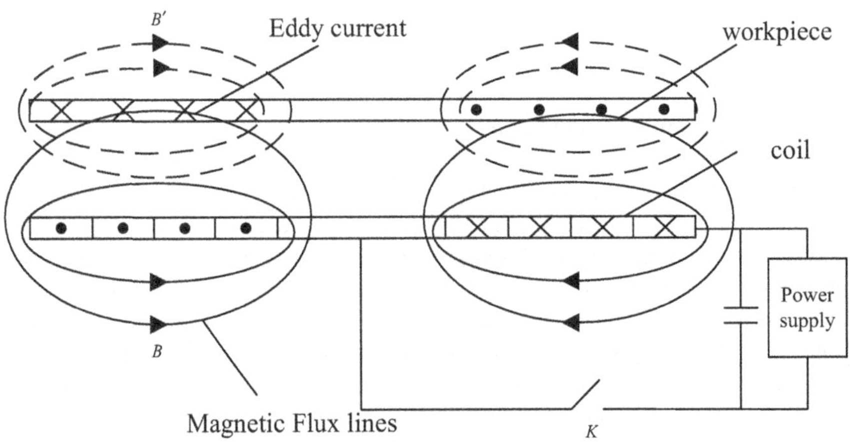

Fig. 1.6 Principle of electromagnetic forming of the plate

In the new century, plastic processing technology is required to be more refined, more economical, and cleaner for high efficiency and quality, low consumption, and low cost. The forming process is expected to be pollution-free. The formed workpiece (blanks) will develop from near-net shaping forming to non-machining shaping forming. The developing cycle of the product should be short. The production process should be able to respond to the market rapidly. In this case, the small energy unit, high efficiency, affined micro-deformation of materials, and high-quality processing make the electromagnetic forming technology a suitable choice. It is foreseeable that electromagnetic forming technology will become one of the most important methods in plastic processing of metals and will be more and more popular in many industrial fields.

The study of electromagnetic forming technology in China began in the 1960s and was interrupted during the Cultural Revolution. In the late 1970s, Harbin Institute of Technology began to study on the basic theories and processes of electromagnetic forming, and successfully developed China's first electromagnetic forming machine for production purposes in 1986 with the help of the experimental equipment. At present, many universities and research institutes in China have carried out many relative researches on electromagnetic forming technology which were applied to real-life productions.

In terms of technology, the strong magnetic field generated by high-power pulsed discharge can be used for metal forming processing, including positive and negative forming of metal workpieces. Because the force transmission at this time is realized by the electronic force in the workpiece, there could be neither mechanical contact nor high temperature. The residual stress of the workpiece is not much therefore the goal of good controllability, repeatability, and high production efficiency can be achieved.

1.3.2 Military Applications

1.3.2.1 Applications of Electromagnetic Emission Technology

The emission technology comes from the advancement of weapon in wars. Electromagnetic emission technology can be divided into the following categories, namely mechanical emission, chemical emission, and electromagnetic emission. Chemical emission can be applied to traditional artillery, rockets, light air guns, thermal transmitters, ram accelerators, etc.

Electric emission technology is an emission technology that uses electric energy to emit objects, and mainly includes electromagnetic emission technology and electrothermal emission technology. Electromagnetic emission technology is an emission technology that uses electromagnetic energy to do work and converts electromagnetic energy into payload kinetic energy such as projectiles. Electrothermal emission technology is an emission technology that uses electric energy to generate plasma (or chemical gas) and emit objects. Compared with the conventional chemical emission method, the electromagnetic emission method has obvious advantages. It can provide massive kinetic energy, which can accelerate payloads such as projectiles to ultra-high initial velocity and emission speed that are difficult to achieve by chemical methods. It's also easier to regulate the speed under high precision, long range, and huge power. The emission process will operate smoothly without being interrupted by noise or smog. It is widely used in controlled nuclear fusion tests and high-voltage physics. It can also be used in the launch of electromagnetic cannonballs and missiles, as well as in electromagnetic aircraft launch systems. In the field of transportation, it is applicable to high-speed electromagnetic trains and pumping units; and it has a very promising growth prospect in the fields of ultra-long-range suppression, air defense and antimissile, and micro satellite launch.

An electromagnetic transmitting device consists of a transmitting device body, a component to be transmitted, and a high-power pulsed power supply. The main body of the transmitting device is the load of the pulsed power supply which provides strong currents. The transmitted component consists of the transmitting payload and its bearing mechanism. The high-power pulsed power supply is the most critical part of the whole device. Pulsed alternator, as a new type of pulsed power supply, integrates the single components in multiple links of energy storage, conversion and regulation. It also simplified the link between power input and power output of railgun load meanwhile optimizing the density of energy and power. Therefore, it is one of the most promising miniaturized railgun pulsed power supply in the future, and it is the first option to realize the dream of ground-based electromagnetic weapon systems in the future.

In the 1970s, the first electromagnetic cannonball (rail gun) of the Australian National University Test Plant in Canberra accelerated the plastic blocks which weighed 3 g to a speed of 6 km/s, exemplifying the potential of electromagnetic emission weapons. Since then, the application of electromagnetic emission technology in the military has become a very popular topic and has begun to make great progress. Currently, the studies in United States are at the advanced level worldwide.

As a developing high-tech weapon, the electromagnetic gun has been valued by navies around the world. It is regarded as a new type of weapon in the future and has a wide range of military applications. Compared to traditional artillery, electromagnetic gun weapons have the following advantages:

1. Huge electromagnetic driving force, and high velocity of projectile

The pulsed power of the electromagnetic launch is about 10 times that of the artillery, here comes the high velocity.

2. Outstanding stability of projectile

The thrust of electromagnetic projectiles in the barrel is called electromagnetic force. It's an affined force and it's easy to control. Therefore, the projectile is quite stable and as a result the accuracy of the projectile can be improved.

3. High-level concealment

The electromagnetic gun does not bring out flames or smoke or shock waves when it is launched, so it is relatively concealed and difficult to find.

4. Adjustable projectile emission energy

The magnitude of the electromagnetic force can be quickly adjusted according to the nature and the range of the target, thereby controlling the projectile's emission energy.

5. More economical

Compared to conventional weapons, artillery propellants require $ 10 per Joule of energy, while electromagnetic guns only require $ 0.1.

There are four categories of electromagnetic guns: rail guns, electric heat guns, coilguns and reconnection guns. Coilguns are similar to rail guns. They both accelerate the shell to a corresponding emission speed with electromagnetic forces before launching. They are also weapons of mass destruction aiming at the large quantities of kinetic energy of the shell. Electrothermal guns often have plasma burners at the tails of the shell. When a high voltage is applied to the electrodes of the plasma burner, an arc will be generated, which makes the plasma overheat and be turned into a high-pressure plasma, in this case accelerates the projectile. The reconnection gun is a multi-stage accelerated non-contact electromagnetic launcher, but it requires a certain initial speed of the shell before entering. It is the latest form of the electromagnetic gun.

High-power pulsed power is an important part of the system of the electromagnetic weapon system. The development of pulsed power is a decisive factor for the advancement of electromagnetic weapons. The development of pulsed power supplies aims at high power density, high energy, and miniaturization. At the same time, the stability and service life cycle of pulsed power supplies under high-frequency repetitive emissions also need to be considered. Figure 1.7 shows a typical electromagnetic gun.

Fig. 1.7 Typical electromagnetic gun

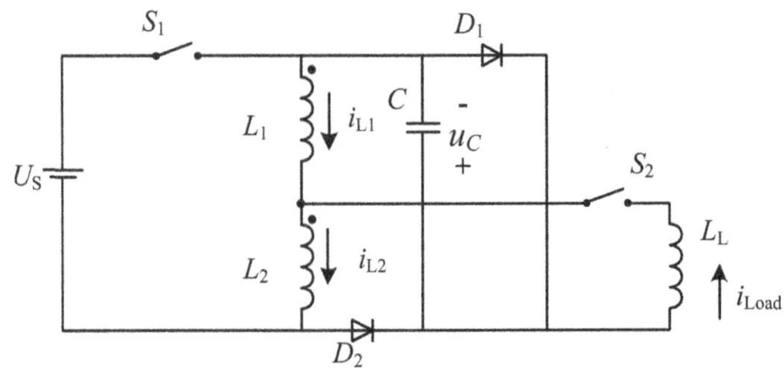

Fig. 1.8 New capacitance transducing inductive energy storage pulsed power system

In order to obtain high-power pulsed currents with amplitudes at mega ampere level and pulse width at millisecond level which are required for electromagnetic emissions, capacitor banks are normally used as pulsed energy sources for intermediate energy storage. However, the density of the capacitor energy storage is relatively low while the volume is large. This is not suitable for a pulsed power supply for the mobile platform of roadbed. Compared with capacitor energy storage, inductors have a higher energy storage density. Therefore, the pulsed power supply using the inductor as an intermediate energy storage has a certain prospect in electromagnetic emission. However, the inductive energy storage electromagnetic emission pulsed power supply puts high requirements for charging power supply, and the main problems, such as high voltage will be generated when the disconnect switch is turned off, need to be solved.

To lower the bars for the charging power supply, a large inductance can be added during the charging process. Circuits such as inductive circuits with series charging and parallel discharging or an inductive energy storage circuit that achieves energy compression with the gradual elimination of mutual inductance between inductors can be taken into consideration. The new capacitance transducing inductive energy storage pulsed power system is as shown in Fig. 1.8. This circuit also uses the transducing capacitor to solve the overvoltage problem of the circuit breaker, without requiring mutual inductance between the coils, and it is easy to achieve series charging and parallel discharging.

Over the past many years, the United States Navy has begun research on large-scale electromagnetic (EM) railguns operating at long distances (over 200 miles) and has made corresponding progress. In order to realize a useful payload range, a diameter of approximately 150 mm and a length of 10–12 m are required. The projectile launching mass is expected to exceed 20 kg with an initial velocity of 2000 m/s and the corresponding muzzle energy of more than 60 MJ. This will require the pulsed power system to provide more than 100 MJ of energy into the rail gun chamber at a time. For a breech fed monorail system, this energy pulse will be more than 5 MA and the voltage will be up to about 20 kV. The typical pulse width is only a few milliseconds, which corresponds to the time for the entire projectile to pass through

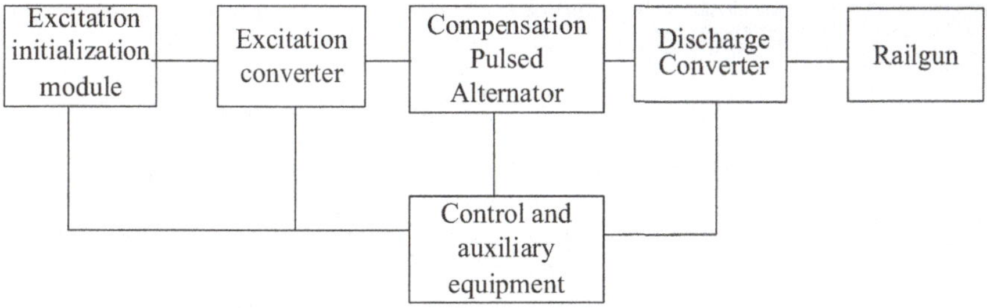

Fig. 1.9 Structure of the railgun system in pulsed alternator

(a) ISL electromagnetic orbit launcher

(b) IAT electromagnetic railgun launcher

Fig. 1.10 Commonly used electromagnetic launcher

the barrel. The ideal pulse shape will lower the acceleration force and acceleration fluctuations from peak to average, while ensuring that there is no significant current interruption before the projectile leaves the muzzle. Figure 1.9 shows a diagram of the railgun system (Fig. 1.10).

1.3.2.2 Applications of Electromagnetic Aircraft Launch Technology

Electromagnetic aircraft launch technology is an electromagnetic technology that uses electromagnetic energy to push the ejected object outward. The main components of an electromagnetic aircraft launch system are a linear synchronous motor, a disc alternator, and a high-power digital cycle inverter.

The electromagnetic aircraft launch system (EMALS), which is used to replace the steam-powered catapult for aircraft, is a research hotspot of the Navy's electromagnetic aircraft launch technology. Figure 1.11 shows a conceptual diagram of an electromagnetic aircraft launch system with a rotating electrical machine that stores hundreds of megajoules of energy as its driving pulsed power supply.

The composition of the UAV electromagnetic aircraft launch system is shown in Fig. 1.12. The drone is accelerated to take-off speed through the electromagnetic

Fig. 1.11 Conceptual diagram of the electromagnetic aircraft launch system and electromagnetic catapult

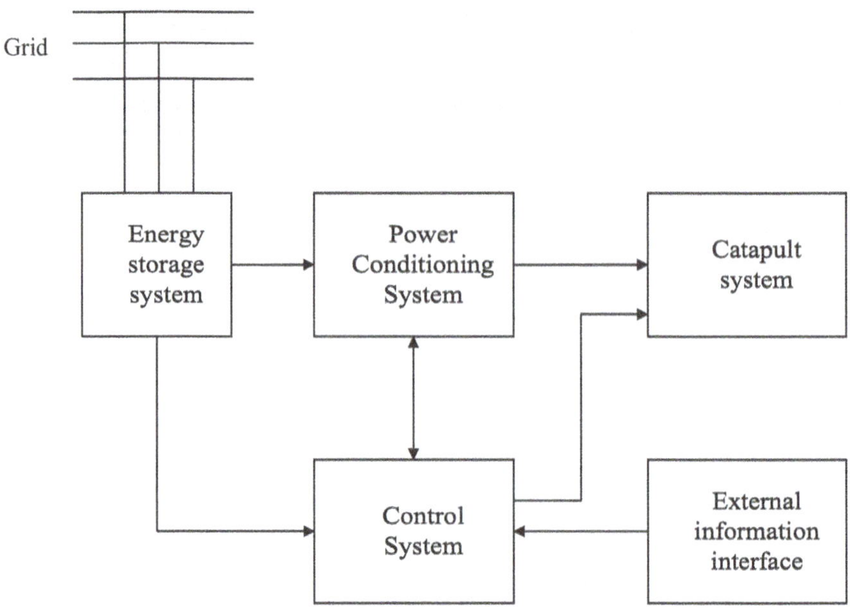

Fig. 1.12 The diagram of the electromagnetic drone launch system

force. The power conditioning sub-system feeds the catapult motor to generate an advancing magnetic field that interacts with the launch system's pulley. Under the effect of electromagnetic force, the pulley carries and accelerates the drone to the required takeoff speed along the ejection stroke. After the drone is disconnected, the pulley returns to the initial launch position at a low speed under the force of the inverse advancing magnetic field.

The Electromagnetic Aircraft Launch System (EMALS) will replace the steam catapult system currently used on aircraft carriers for ejection. With new power electronics technology, the high controllability, high efficiency, and high performance required by EMALS can be achieved. The electromagnetic aircraft launch system is a fully integrated system consisting of an energy storage system, a power electronics system, a linear launch motor, and a control system. These sub-systems are combined to form a high-performance launch system, which greatly enriches the operational capabilities of aircraft carriers in the future. The aircraft carriers will not only be able to carry heavier and faster aircraft, but also allow the ejection of small and light aircrafts that do not match with the existing steam catapults. The development and application of such launch systems can both leverage and enhance existing civilian technologies.

1.3.2.3 Other Military Applications

Ships are considered to be threats for detection of and confronting to the ballistic missiles. Most ships are equipped with radars that can detect regional targets from a few square meters to hundreds of square miles. However, these large-scale explorations require the high-power power supplies whose expected average output powers and peak powers are one megajoules and several megajoules respectively. High electrical power qualities need to be very high therefore the radars are very expensive. One possible alternative is to design a system to support the radar's highest average power requirements in order to address the tracking mode with highest power and lower frequency as well as the pulse power system. This can significantly reduce the cost and size of the power supply.

Pumped high-powered laser weapons is a modern new-concept high-powered laser weapon that uses electric-pulsed power pumps. This strong laser mainly includes electric pumped CO_2 laser, excimer high-power laser, free electron laser (amplification and oscillator modes), and soft X-ray laser. Excitation of strong electromagnetic pulses and strong pulsed electromagnetic fields can be very applicable. On the one hand, they can be used as electromagnetic pulsed weapons for soft and hard munitions; on the other hand, they can generate strong electricity.

Magnetic pulses are used in various simulation tests to strengthen their own electronics and power systems meanwhile avoiding electromagnetic interferences and damages.

There are two types of excitations of high-power electromagnetic pulses: one is to use high-power particle beams to generate high-power microwaves for microwave weapons; the other is to excite suitable loads directly with high-power pulsed power to generate various broad-spectrum electromagnetic pulses. Electromagnetic pulses are mainly used to: (1) make high-power microwave weapons (including microwave "cannons" and microwave bombs); (2) make electromagnetic "missiles"; (3) make electromagnetic pulsed bombs; (4) simulate electromagnetic pulses.

1.4 Different Kinds of High-Power Pulsed Power Supplies

1.4.1 Capacitive Energy Storage Pulsed Power Supply

Capacitive energy storage is also called electric field energy storage. The main energy storage elements are pulsed capacitors. Figure 1.13 shows a schematic diagram of a capacitor energy storage pulse power supply. When switch S_1 is closed, the DC power source E_0 charges the storage capacitor C through the resistor R_1 directly until the voltage across the capacitor is equal to U. R_c is the insulation resistance of the capacitor. When the capacitor discharges by itself, the time constant $\tau_C = R_c C$. When switch S_2 is closed, the capacitor C releases the stored energy to the load Z_L within milliseconds or microseconds, so that a high-power pulse current can be generated on the load.

The energy stored in the capacitor E_1 is

$$E_1 = \frac{1}{2}CU^2 \tag{1.1}$$

It can be known from formula (1.1) that the energy stored in the capacitor is proportional to the square of the voltage at the end of the capacitor and the capacitance value.

For parallel capacitors, the volume is: $V_1 = Sd$, where S, d represent the area of the capacitor plate and the distance between the parallel plates. If ignoring the remaining volume, the volume power density of the capacitor is

$$\kappa_V = \frac{E_1}{V_1} = \frac{\frac{1}{2}\frac{\varepsilon S}{d}(Ed)^2}{Sd} = \frac{1}{2}\varepsilon E^2 = \frac{1}{2}\varepsilon_0\varepsilon_r E^2 \tag{1.2}$$

where E is the electric field strength; ε is the dielectric constant; $\varepsilon = \varepsilon_0\varepsilon_r$; ε_0 is the vacuum dielectric constant; $\varepsilon_0 = (36\pi \times 10^9)^{-1}$ F/m; ε_r is the relative dielectric constant and the unit of $\kappa_V$is J/cm^3.

With the discharge current in Fig. 1.13, millisecond or microsecond pulsed currents can be obtained. The cycle of the discharging pulse is:

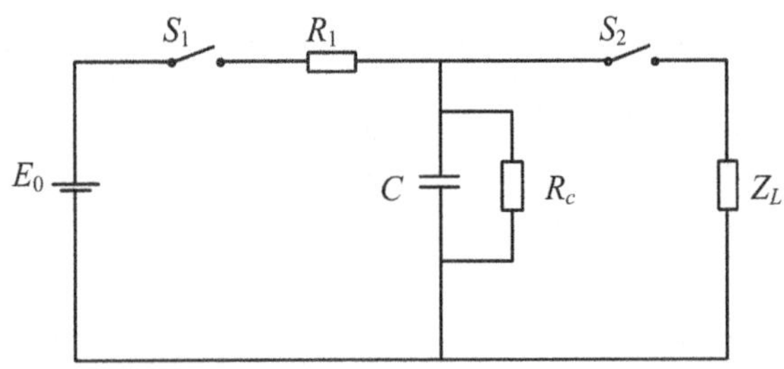

Fig. 1.13 Schematic diagram of capacitive energy storage and its discharging

(a) Thunderbolt 32 MJ banks of capacitors

(b) U.S. Navy's 100 MJ banks of capacitors

Fig. 1.14 Energy storage capacitor bank

$$T = 2\pi\sqrt{LC} \tag{1.3}$$

L is the total inductance of the discharge loop. If a nanosecond pulsed current is required, the value of the total inductance of the circuit needs to be further reduced, but it is difficult to achieve in the actual process. Therefore, the nanosecond pulsed current is expected to adopt a new circuit topology. For example, the capacitor can be charged with a Marx generator to charge the transmission line. When the transmission line is discharging to the load, a strong pulse current of the nanosecond level can be obtained. Figure 1.14 is an energy storage capacitor group.

Pulsed capacitors are generally used in high-voltage test devices such as pulsed voltage generators, pulsed current generators, and oscillating circuits. In addition, they can also be used in electromagnetic forming, hydroelectric forming, hydroelectric breaking, energy storage welding, subsea exploration, production of high-temperature plasma ultra-strong magnetic field, light sources with strong impact, lasers and other devices.

High-energy-storage-density pulsed capacitors are now widely used in pulsed power supplies, medical devices, electromagnetic weapons, particle accelerators and environmental protection. The energy storage pulsed capacitors have gone through the development of paper/aluminum foil structure, paper film structure, and metalized electrode structure. Their energy storage densities have increased from dozens of J/L to nearly kJ/L (2 ~ 3 kJ/L for laboratory samples).

Each significant increase in capacitor energy storage density can be traced back to the application of new materials or processes. The impregnating agent ranges from the mineral oil of paper dielectric capacitors to the rapeseed oil of metallized film capacitors, and the electrodes range from aluminum foil to metallized vapor deposition then to split metallized electrodes. The medium developed from paper to paper film to full film then to composite membrane. The new dielectric film will be the focus in the process of developing the next generation of energy storage capacitors. Table 1.4 is a list of some commonly used pulsed capacitor products.

Table 1.4 Commonly used pulsed capacitor products

Model	Rated voltage (kV)	Nominal capacitance (μF)	Scope of application	Mass (kg)	Dimensions (length × Width × Height) mm
MW1-500	1	500	Divided low-pressure arm	29	380 × 110 × 855
MW3-200	3	200	Divided low-pressure arm	23	372 × 143 × 480
MW4-50	4	50	Electric shock	22	302 × 122 × 405
MWF50-3	50	3	Continuous pulse	146	400 × 328 × 930
MWF30-18	30	18	Oscillating circuit	220	310 × 310 × 1540
MWF40-0.03	40	0.03	Rectification filtering	8	Φ185 × 210
MWF220-0.25	–	–	Impulse voltage	320	Φ630 × 850

The Marx generator is a high-voltage pulsed device that uses capacitors in parallel to charge and then discharge in series. Marx proposed in 1924 that it could imitate the processes of lightning and operating overvoltage, so it is often used for insulated impulse withstand voltage, dielectric impulse breakdown, discharging and other tests. Figure 1.15 shows the Marx generator device.

The principle of the Marx generator is very simple. Figure 1.16 shows the schematic diagram of the inductive isolated Marx generator. In the following figure, R_0 is the charging matching resistor; S is the charging spark switch; G_1, G_2, G_n, G_L are discharging spark switches. L_0 is the isolated inductor; C_0 represents the capacitors at each level; R_L is the load. Compared with the traditional Marx generator, it's featured with an Inductor L_0 which is used as a substitute for resistor R between

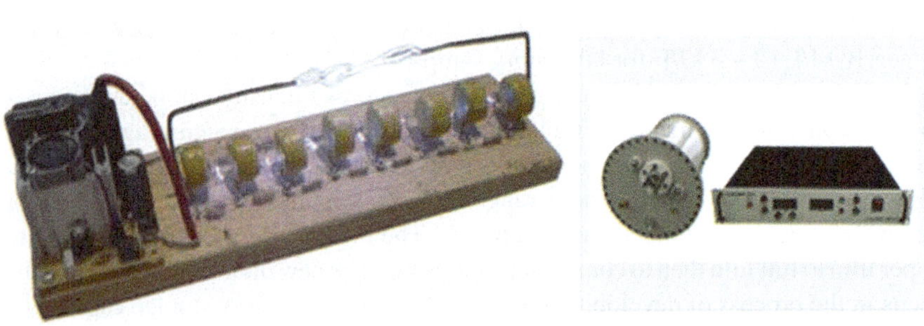

Fig. 1.15 Marx generator

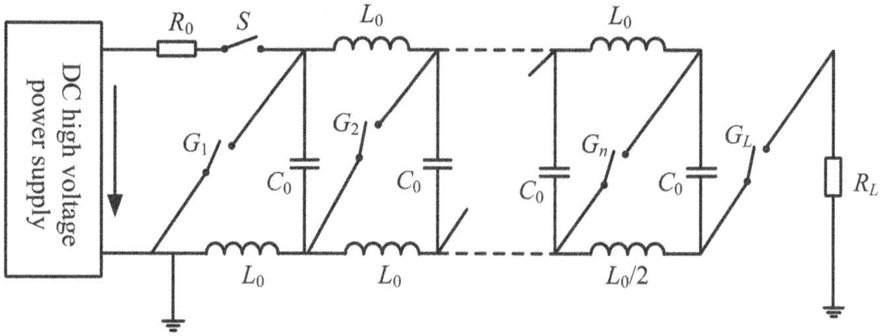

Fig. 1.16 Schematic diagram of the inductive isolated Marx generator

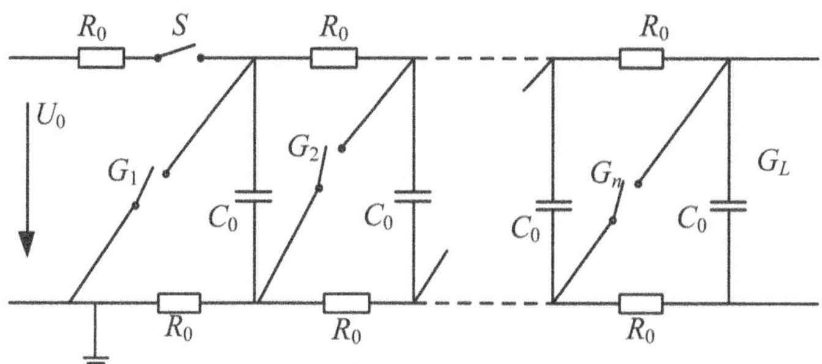

Fig. 1.17 The circuit diagram of the Marx generator

the capacitors C_0 of each level. It also has the functions of charging conduction and discharging isolation. Furthermore, an additional charging spark switch *S* is objected.

Figure 1.17 shows a schematic diagram of a traditional Marx generator. The difference from Fig. 1.16 is that the inductor is changed into a resistor. The principle behind is that the capacitors are charged in parallel, and then the capacitors at each level are discharged through the switch. Connected in series, the voltage equals to the sum of the terminal voltages of a capacitor, and at the same time, the load is discharged to generate a nanosecond high pulsed current.

1.4.2 Inductive Energy Storage Pulsed Power Supply

Inductive energy storage pulsed power supply is essentially a magnetic-field energy storage pulsed power supply, in which energy is stored in the magnetic field of the coil. It is released to the load during discharging for a strong pulsed current.

The advantages of inductive energy storage systems are: (1) high energy storage density, small size, and low cost; (2) it can generate extremely high pulsed power (at

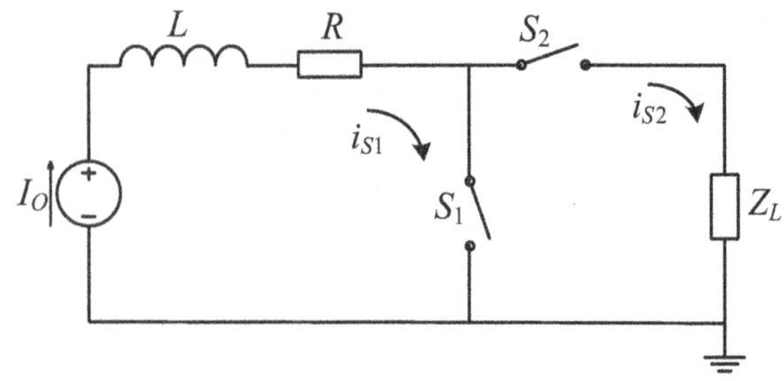

Fig. 1.18 Schematic diagram of the inductive energy storage

the level of microseconds or sub-microseconds). Disadvantages are as follows: (1) the technology of circuit breaker is challenging. It is not as simple as short-circuit switch; (2) It is not very efficient to transfer energy to the load.

The schematic diagram of the inductive energy storage is shown in Fig. 1.18. When the switch S is closed and S_2 is open, the power supply charges the inductor L and stores the power of the power supply in form of a magnetic field in the inductor coil. When the switch S_1 is open while S_2 is closed, the loop voltage equals to the sum of the voltage of the inductor and the power supply, and at the same time discharges to the load Z_L to generate a strong pulsed current.

During the conversion (commutation) of charging and discharging, it can be known from the circuit theory that the total magnetic flux of the energy storage inductor remains stable. The charging time constant of the inductor is $\tau = L/R$, and the magnitude of τ is generally seconds. The time of discharging process is related to the parameters of the load Z_L, which will not be discussed here.

The energy storage E_2 of the inductor can be expressed as:

$$E_2 = \frac{1}{2}Li^2 \tag{1.4}$$

The volume of the coil is: $V_{\mathrm{L}} = Sl$, where S is the cross-sectional area of the coil, and l is the length of the coil. Therefore, the inductive energy storage density can be:

$$\kappa_L = \frac{E_2}{V_L} = \frac{0.5Li^2}{Sl} = \frac{\frac{B^2 Sl}{2\mu}}{Sl} = \frac{B^2}{2\mu_0\mu_r} \tag{1.5}$$

In this formula, B is the magnetic induction strength, μ is the magnetic permeability, μ_0 is the vacuum permeability, and μ_r is the relative permeability.

From formula (1.5) it can be seen that the energy storage density of the inductor is inversely proportional to the magnetic permeability and proportional to the square of the magnetic flux density (Fig. 1.19).

With the progress of new concepts such as electromagnetic launches and controlled nuclear fusion, more and more attention has been paid to high-power

Fig. 1.19 Inductor mounted to cryostat

and high-repetition-frequency continuous pulse technology. The research and development of repetitive disconnect switch technology is the key point to bring repetitive pulse into use.

1.4.3 Chemical Energy Storage Pulsed Power Supply

Chemical energy is widely used in the field of pulsed power because of its high energy storage density. At present, chemical energy is mainly used for chemical pulsed power, chemical energy magnetic flux compression, shrinkage generator and pulsed magnetic fluid generator. Chemical sources can be divided into alkaline batteries, acid batteries, neutral batteries and machine electrolyte battery.

According to method of storage and nature of work, it can also be divided into: (1) primary batteries such as zinc-manganese dry batteries, zinc-mercury dry batteries, zinc-silver dry batteries, and lithium batteries; and (2) secondary batteries including lead-acid batteries, nickel–cadmium batteries, zinc-air batteries, and lithium-iron sulfide batteries. (3) Reserve batteries, such as magnesium-silver batteries, lead-perchloric acid batteries, and thermal batteries. (4) Fuel cells like hydrogen–oxygen fuel cells and hydrazine-air fuel cells.

1.4.3.1 Battery

A storage battery is also called a secondary battery. It is a device that converts chemical energy into electrical energy. Electrical energy is converted into chemical energy and stored before use. When working, chemical energy is converted into electrical energy, and it can be charged and used multiple times. According to the

Table 1.5 Commonly used batteries

Battery type	Energy density/(Wh/kg)	Power density/(W/kg)	Working lifespan
Ferronickel	20 ~ 45	65 ~ 90	2000 ~ 5000
Nickel–cadmium	25 ~ 45	200 ~ 500	1000 ~ 2000
Silver–zinc	50 ~ 150	200 ~ 400	100 ~ 2000
Nickel–zinc	60 ~ 70	100 ~ 200	200 ~ 300
Lead acid	20 ~ 35	20 ~ 175	300 ~ 2000

type of electrolyte, it is divided into acid battery and alkaline battery. "The battery" usually refers to lead-acid battery. The alkaline battery electrolyte mainly consists of potassium hydroxide, such as cadmium nickel batteries, nickel-hydrogen batteries, etc., while the acid battery electrolyte is sulfuric acid aqueous solution, taken lead-acid as an example. Table 1.5 show the commonly used batteries.

Batteries can be used in pulsed power devices, which can obtain strong pulsed currents in seconds and milliseconds. The discharge circuit is shown in Fig. 1.20. After the switches S_1 and S_2 are closed, the load current generated by them is:

$$i(t) = \frac{U}{R_i + R_c}\left(1 - e^{\frac{t}{\tau}}\right) \tag{1.6}$$

The time constant $\tau = L/R$, R_i and R_c refer to the internal resistance and the external battery resistance of the battery respectively. The current in the circuit can reach 10 kA within one second. Lead-acid batteries have high and stable operating voltages, a wide range of temperature and current changes, and are easy to manufacture and low in cost. Therefore, they are still widely used in rechargeable and reused electrochemical energy storage devices.

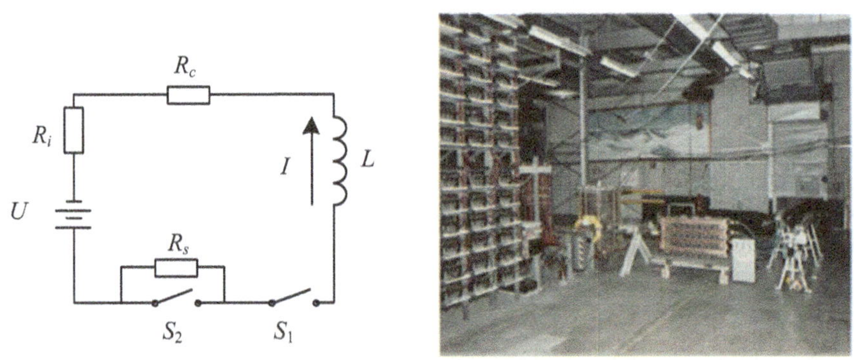

(a) Battery discharge circuit to inductive load

(b) Power supply of lead-acid battery pack

Fig. 1.20 Battery discharging circuit and lead-acid battery pack

1.4.3.2 Flux Compression Generator

Magnetic Flux Compression Generator (MFCG), also known as Flux Compression Generator (FCG). It uses explosives to drive the conductor armature to rapidly compress the stator winding for the generation of the magnetic flux in the cavity, so that it accumulates in a small volume to form an ultra-high magnetic field, or amplifies the current (and electromagnetic energy) in the load connected to the stator and armature. The principle of the magnetic compression generator was first proposed by the Soviet academician Sakharov in the 1950s. F. Chemyshev and C. M. represented by Fowler, the Russian and American scientists have made great progress in nuclear weapons after decades of hard work, which has been widely used in electromagnetic emission, poly-explosive plasma, strong pulsed battery radiation and strong magnetic fields.

According to the winding structure, the magnetic flux compression generator (MFCG) can be classified into three categories: single turn (current loop), coil and solenoid. The main body of the explosive spiral winding MFCG is a spiral winding and an armature fixed coaxially at the inner end of the winding. The armature is a cylinder made of high-conductivity, high-ductile material (such as aluminum), which is filled with explosives. The inner diameter of the winding is about twice the outer diameter.

In 1942, Alvin stated that the magnetic lines of flux would be "frozen" in the fluid in a completely conductive fluid. This is a basic theory in MHD (Magnetohydrodynamics), which is called the magnetic freezing. The concept of magnetic freezing is also applicable to solid conductors with good conductivity. The closed-surface magnetic flux is conserved in any of the ideal conductive fluids in motion. The principle is shown in Fig. 1.21.

The initial magnetic flux in a conductor with a cross-sectional area of A is $\Phi_0 = B_0A$. At this time, the magnetic energy in the cylinder with an energy storage volume of V_0 is: B_0

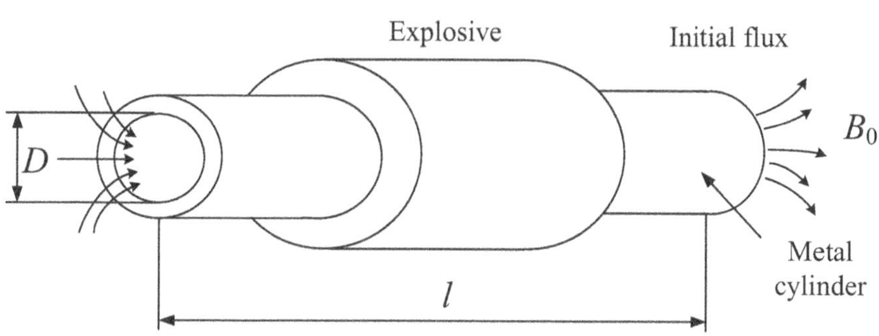

Fig. 1.21 Schematic of magnetic freezing

$$E_0 = V_0 \frac{B_0^2}{2\mu_0} = \frac{\pi D^2 l}{4} \cdot \frac{B_0^2}{2\mu_0} \tag{1.7}$$

where D is the diameter of the barrel before the explosion. When the explosive explodes, the diameter of the cylinder is compressed to d. At this time, the following can be deducted from the law of conservation of magnetic flux:

$$\Phi_0 = \int_S B \cdot dA = \text{const} \tag{1.8}$$

Given the magnetic flux remains unchanged, the magnetic induction strength is

$$B = B_0 \left(\frac{D}{d}\right)^2 \tag{1.9}$$

The magnetic energy stored in the compressed volume V at this time is:

$$E = V \frac{B^2}{2\mu_0} = \frac{\pi d^2 l}{4} \cdot \frac{B_0^2}{2\mu_0} \left(\frac{D}{d}\right)^4 = E_0 \left(\frac{D}{d}\right)^2 \tag{1.10}$$

The work done by an external force to compress the magnetic flux is converted into the energy of magnetic field magnetic field, which is proportional to the square of the diameter. Before and after compression, the magnetic flux is conserved. It can be expressed as $\Phi_0 = L_0 I_0 = LI$ (L_0 and I_0 represent the initial inductance and current, and L and$I$ represent the inductance and current after compression).

As for the magnetic flux compression generator, it can be equivalent to the basic circuit diagram in Fig. 1.22, where the generator consists of three parts namely the initial magnetic flux power supply, the body of the generator, and the load. The initial current I_0 of the initial magnetic flux power is generated by the discharging of energy storage capacitor. After that, an initial magnetic flux Φ_0 is established in the loop. In the following figure, L_g is the variable linear inductance, L_0 is the distributed inductance caused by the lead, L_L is the load inductance, and R is the generator

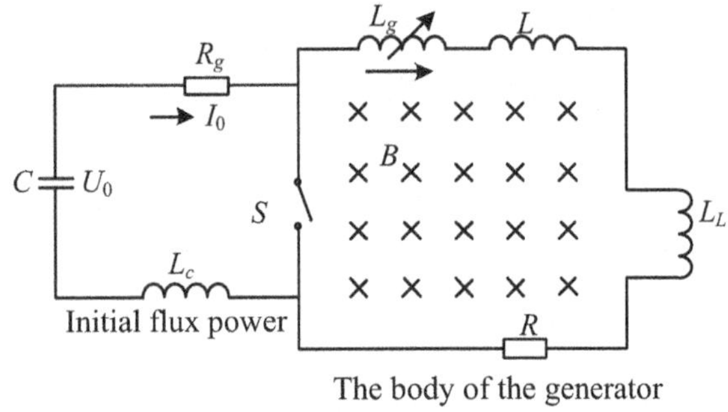

Fig. 1.22 Equivalent circuit of magnetic flux compression generator

resistance. When I_0 reaches its maximum value, switch S is closed (equivalent to the circumstance in which the explosive drives circuit conductor or armature to close the circuit). As the explosive pressure forces $L_g(t)$ to start to decrease from L_0, the current gradually increases, and the load gains increased energy. Due to the low internal impedance of the MFCG, it loads multiple low-impedance coils, and pulse transformers are required for high-impedance use.

Assuming when $t = 0$, resistance $R = 0$, and $L_g(0) = L_0$. According to energy conservation law, the relationship among load current $I_L(t)$, distributed inductance L_0, and magnetic field energy storage $W_L(t)$ on load inductance is as follows.

The relationship between load current I_L(t) and magnetic field energy storage $W_L(t)$ is as follows:

$$\left.\begin{aligned} I_L(t) &= (L_0 + L + L_L)I_0/(L_g(t) + L_0 + L_L) \\ W_L(t) &= (L_0 + L + L_L)W_0/(L_g(t) + L_0 + L_L) \end{aligned}\right\} \quad (1.11)$$

$$W_0 = (L_0 + L + L_L)I_0^2/2$$

where W_0 is the initial magnetic energy. After explosion, $L_g = 0$, final energy of the final currents If, L_0 and L_L can be expressed as:

$$\left.\begin{aligned} I_f &= (L_0 + L + L_L)I_0/(L + L_L) \\ W_f &= (L_0 + L + L_L)W_0/(L + L_L) \end{aligned}\right\} \quad (1.12)$$

In an actual circuit, R varies with the skin depth of the conductor, the temperature, and the length of the loop during the explosion. R is a complex multivariate function. Supposing R is a constant during the explosion, and Lg is the only function of time. The differential circuit and initial conditions of the circuit at this time are:

$$\frac{d}{dt}\left[(L_g(t) + L + L_L)I(t)\right] + I(t)R = 0,\ I(0) = I_0 \quad (1.13)$$

The loop current is I_0 when time $t = 0$, and the initial inductance of the generator is L_g (0) = L_0; in this case the solution $I(t)$ of Eq. (1.13) and the energy of the system at different times $W(t)$ are

$$\left.\begin{aligned} I(t) &= I_0\frac{L_T(0)}{L_T(t)}\exp\left[-\int_0^t \frac{R}{L_T(y)}\mathrm{d}y\right] \\ W(t) &= W_0\frac{L_T(0)}{L_T(t)}\exp\left[-2\int_0^t \frac{R}{L_T(y)}\mathrm{d}y\right] \end{aligned}\right\} \quad (1.14)$$

In the formula, the total loop inductance $L_T(t) = L_g(t) + L + L_L$.

Table 1.6 The classification and performance of the typical motors for pulsed power

Types	Energy density (kJ/kg)	Power density (kW/kg)	Typical pulse width (s)	Typical voltage (V)	Short-circuit current (kA)	Energy storage time (s)	Specific energy density (MJ/m^3)
DC pulsed generator	0.32	0.3	1	1800	1	100	20
Unipolar generator	8.5	70	0.1 ~ 0.5	100	2000	415	150
Synchronous generator	1.3	0.7	71	6900	6	3000	30
Compensated pulsed alternator	3.8	250	10^{-4} ~ 10^{-3}	6000	71	254	100

1.4.4 Inertial Energy Storage Pulsed Power Supply

The inertial energy storage device stores energy in a moving object, which is not only convenient for storage, but also only needs to sharply decelerate the moving object to release the stored energy. Generally, a low-power driving mechanism slowly accelerates a rotor or flywheel of a certain mass for a relatively long time to make it rotate to obtain sufficient kinetic energy for storage, and then uses inertial pulses to drive the power generation equipment to convert mechanical energy into pulse-electromagnetic energy. Two common examples are pulse generators and unipolar generators. Their advantages are high energy storage density, compact structure, low cost, and high mobility. They are mainly used in the power of electromagnetic transmitters, synchrotrons, sintered metal powders, electromagnetic spraying, simulated seismic pulses, and pulsed metal forming. The typical rotary electric machines for pulse power are shown in Table 1.6.

1.4.4.1 DC Pulsed Generator

Unlike a single-pole generator, a DC pulse generator only has a single-turn rotor. It can be wound with a wire to make a multi-turn rotor, and it is easy to obtain high voltages in the kilovolt range.

The DC generator is composed of an excitation magnetic field, a rotor armature and an end commutator, and a DC voltage drawn from the brush. To obtain high-energy pulses, the method of flywheel inertial energy storage should be used. Rotors of generators and flywheels of larger mass are often driven by asynchronous machines for a gradually increasing amount of kinetic energy. When it has reached the rated speed, an excitation magnetic field is provided to the generator, and the motor generates an unloaded voltage. After the generator becomes stable with the external circuit connected at the same time. It can discharge to the load and generate a high power

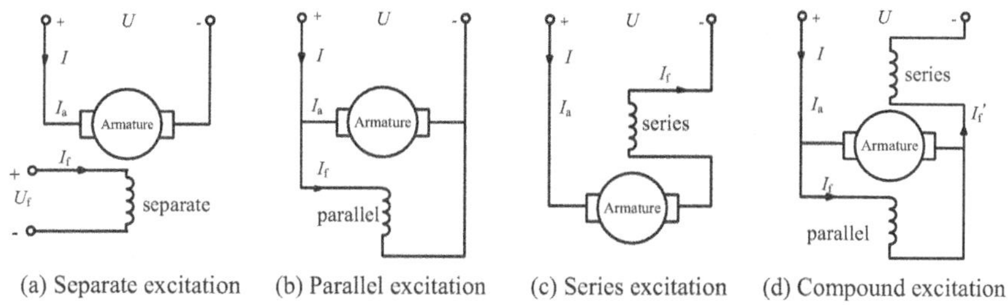

Fig. 1.23 Wiring diagram of the four excitation methods of DC generator

and strong pulsed current. The generator is decelerated, and the mechanical energy stored in the rotor and the flywheel is converted into electromagnetic energy in the form of pulses.

The excitation winding must be cut off during the no-load start of the generator and the acceleration of the flywheel. The generator can be operated by a single motor, or a set of engines consisting of multiple paralleled motors. The magnetic field is the medium through which the motor carries out the conversion of electrical energy. There are two types of excitation methods for DC motors: one is the permanent magnet magnetic field, which is used in some special micro-motors; the other is the electromagnet magnetic field, which is generated by the current flowing into the exciting winding on the core of the main pole, which is called the excitation magnetic field. According to the different excitation methods, they can be categorized as follows.

1. Separately excited DC motor

For separately excited DC motors, the excitation winding is provided by other DC power sources separately. As shown in Fig. 1.23a, U is the armature voltage, U_f is the excitation voltage, I_a is the armature current, and I_f and $I_{f'}$ are the excitation currents. I is the main supply current, and the motor current direction is shown in the figure.

2. Parallel excitation DC motor

As shown in Fig. 1.23b, it can be seen that the excitation winding and the armature winding are connected in parallel, and the armature voltage equals to the excitation voltage. For a parallel-excited generator to self-excite and generate voltage, there must be remanence in the magnetic circuit of the motor. When the direction of the magnetomotive force generated by the excitation current is the same as the direction of the remanence, they are correlated positively. When the two directions are opposite, negative feedback is formed.

In negative feedback, the remanent magnetic field is suppressed and the voltage cannot be established; in positive feedback, the breath magnetic field is strengthened, which increases the armature-induced electromotive force, thereby further strengthening the excitation current and the air gap magnetic field.

3. Series excitation DC motor

The field winding is connected in series with the armature winding, and the armature current is the field current. This can be shown in Fig. 1.23c.

4. Compound excitation DC motor

The excitation winding is divided into two parts, one is connected in series with the armature winding, and the other one is connected in parallel with the armature winding, as shown in Fig. 1.23d. If the directions of the two magnetomotive force are the same to each other, this is called cumulatively compounded excitation; if the two directions are opposite to each other, it is called differentially compounded excitation. Commonly used compound-excitation generators are cumulative ones, which can be divided into three further categories namely flat compounding excitation, under compound excitation, and over compound excitation.

For a DC generator, the effective length of the armature conductor is l and the linear velocity of the conductor cutting the air-gap magnetic field is v. In this case the induced electromotive force of each conductor is

$$e_c = b_\delta l v \tag{1.15}$$

If the total number of armature conductors is Z_a and the number of each branch in series is $Z_a/(2a_=)$, then the electromotive force E_a is

$$E_a = \sum_{1}^{Z_a/2a_=} b_\delta l v = l v \sum_{1}^{Z_a/2a_=} b_\delta(x_i) \tag{1.16}$$

Then the linear velocity of the rotor should be considered as $v = 2p\tau\frac{n}{60}$, where τ is polar distance, $2p\tau$ is the armature perimeter, therefore

$$E_a = 2\frac{pn}{60}\frac{Z_a}{2a_=}(B_{av}l\tau) = \frac{pZ_a}{60a_=}n\Phi \tag{1.17}$$

Similarly, the electromagnetic torque of the DC motor is

$$T_e = Z_a B_{av} l\left(\frac{I_a}{2a_=}\right)\frac{p\tau}{\pi} = \frac{pZ_a}{2\pi a_=}\Phi I_a \tag{1.18}$$

1.4.4.2 Homopolar Generators

Homopolar generators have been used in the field of pulsed power for many years and are also called pulsed unipolar generators. The conductive disc (rotor) itself can store energy. It rotates in an external axial magnetic field. Because the disc cuts the magnetic lineation, an induced electromotive force is generated between the outer edge of the disc and the shaft, and it is led to the external circuit load by a brush.

The induced electromotive force is

$$V = \frac{1}{2}Br^2\omega \tag{1.19}$$

where B is the magnetic induction intensity, r is the radius of the rotor, and ω is the rotational angular velocity of the motor. Usually energy storage refers to the inertial energy storage of the rotor. Because the rotor of a unipolar generator is both an energy storage body and a single-turn coil that induces electromotive force, the output voltage is generally low, but its low internal resistance (<10μ) serves as a makeup for the shortcomings of low voltage.

The single generator is suitable as a pulse power source, especially the energy storage required is dozens to hundreds of MJ. The excitation coil used by the single pole generator is relatively simple and the rotor has no windings, so the rotor can be accelerated to high speed quickly. Also the possibility to be converted into electrical energy in milliseconds exists. However, the inherent equivalent capacitance of unipolar generators is very large (which can be thousands of farads), so it's a bit inconvenient to use the inductive load.

In excitation field with constant current, its equivalent capacitance is

$$C_e = \frac{2W_k}{V^2} \tag{1.20}$$

where W_k is the kinetic energy of the rotor, expressed as the moment of inertia J and the magnetic flux Φ as

$$C_e = \frac{4\pi^2 J}{\Phi^2} \tag{1.21}$$

Methods to reduce the influence of the capacitance of a unipolar generator: (1) a self-excited unipolar generator powered by its own magnetic field coil, at which time the magnetic field coil functions both as an excitation and an energy storage for the inductor; (2) It is more advantageous to use several unipolar generators in series, especially when adjacent rotors act together as an excitation coil, which not only reduces the equivalent capacitance but also increases the output voltage. Compensated pulsed alternators and unipolar restaurants are more suitable for resistive and capacitive loads, which are not suitable for excitation inductive loads. Figure 1.24 shows a multi-phase homopolar generator system.

(a) 40 MW multi-phase homopolar generator

(b) Westinghouse's 12 MJ homopolar generator

Fig. 1.24 40 MW homopolar generator and 12 MJ homopolar generator system

1.4.4.3 Synchronous Generators

Alternators are divided into synchronous generators and asynchronous generators. Synchronous generators refer to alternators in which there is a fixed relationship between the speed of the rotor n ($n = n_s = 60f/p$) and the frequency f of the power grid. n_s is called synchronous speed. If the frequency of the power grid remains constant, the speed of the synchronous motor during steady state operation is constant regardless of the load, and the non-salient-pole motor according to the shape of the main pole. Figure 1.25 shows the synchronous generator.

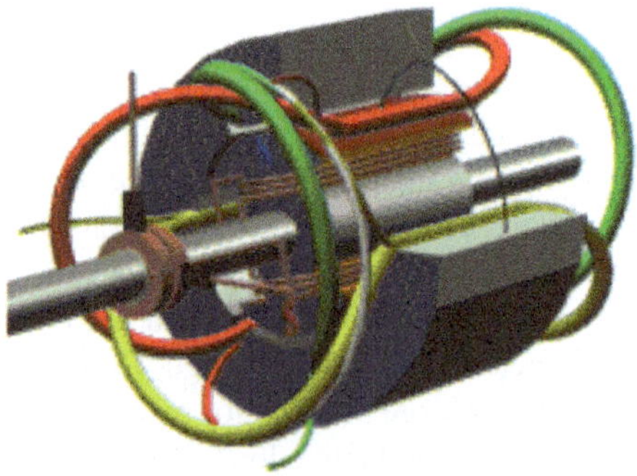

(a) Schematic diagram of synchronous generator

(b) Three-phase synchronous generator

Fig. 1.25 Synchronous generator

According to the structure of the motor, the synchronous generator can be divided into pivot type and transition type. The transition type includes the salient pole motor and the non-salient pole motor.

1.4.4.4 Pulsed Alternators

Pulsed alternator (PA) is an improved generator that can generate larger pulsed power based on the alternator. On the basis of the law of electromagnetic induction and the principle of magnetic flux compression, pulsed alternators integrate energy storage, electromechanical energy conversion and pulse forming together. Compared with the traditional pulsed power system, it has the advantages of high energy storage density and power density, high stability and reliability; and in continuous pulse operation, because of its large moment of inertia, the speed does not decrease too much, so there is no need for flywheel to keep balance. It is therefore often used as a power source for pulsed power systems. The working principle of the pulsed alternator set is shown in Fig. 1.26.

W. F. Weldon et al. from The University of Texas at Austin-Center for Electromechanics (UT-CEM) in 1978 invented the Compensated Pulsed Alternator-CPA. It is a patent based on the alternators with an improvement of the compensation structure. In the same year, the first prototype was developed to verify that the compensation structure can be significantly reduced, and the pulse output power can be increased. Figure 1.27 shows a pulsed alternator set.

There are three methods of compensation of pulsed alternator: active compensation, passive compensation, and selective and passive compensation. Three different compensation methods meet the different requirements of different loads for the current waveform. According to the excitation method, it can be divided into permanent magnetic excitation, electric excitation and hybrid excitation. Compared with

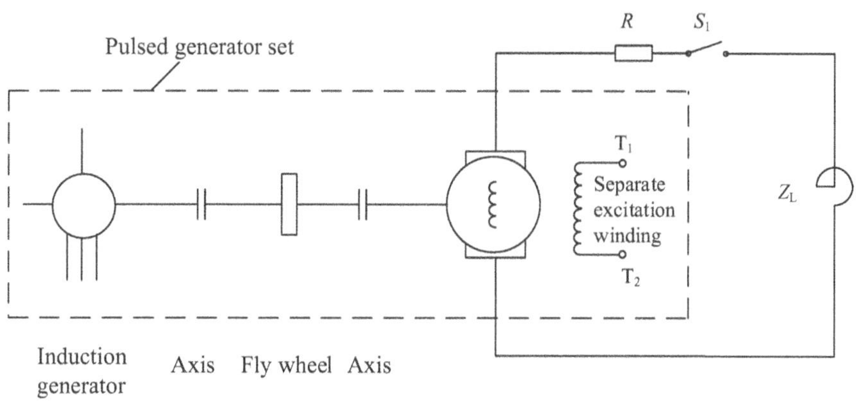

Fig. 1.26 Working principle of pulsed alternator set

Fig. 1.27 J-TEXT Tokamak 100 MW pulsed alternator set

the above-mentioned inertial energy storage pulsed units, CPA has the following comprehensive advantages:

1. It is directly coupled to the initial power source, and integrates inertial energy storage, electromechanical energy conversion and pulse shaping. The motor system has a simple structure, high power density and high energy density.
2. Pulsed shaping is easy to match the load. CPA pulse width is from microsecond level to millisecond level, and it is easy to adjust.
3. CPA provides continuous pulses and is suitable for continuous weapon delivery systems.
4. The current pulse is naturally more than zero and has a "self-switching" characteristic without the need for complicated switching technology.

1.5 Future Developments in Pulsed Power Technology

After more than half a century of development, pulsed power technology has gradually extended from the fields of high-tech and national defense to the industrial and civilian ones. As an important part of the contemporary high-tech field, its development and application are closely related to the development of other disciplines. Analysis of the current trend of pulsed power technology can be summarized as follows:

1. From single pulse to repeated ones with high average power. In the past, pulsed power technology was mainly used for national defense and scientific researches, and most of them were once-run operation. However, industrial and civilian pulsed power technology required a certain average power and must be operated at repeated frequencies.

2. Energy storage technology—aiming at the invention of power supplies with high energy storage density. In many applications, the size and weight of the pulsed power system are decisive factors, such as aircraft detection of underwater objects, shipborne electromagnetic guns, etc., which all require a large pulsed power, and a system which is not too large or bulky. Therefore, the development of pulsed power generators with high energy storage density is one of the main research topics at present.
3. Switching Technology—aiming at the exploration of new high-power switches and high-repetition-frequency switches. The parameters of the switching elements directly affect the performance of the entire pulsed power system, which is a key technology in the pulsed power technology. The US Air Force weapon scientists believe that the current high-power switching technology includes the following aspects: short-pulsed technology, synchronization technology, high-repetition-frequency technology, and long-life technology. The difficulty lies in high-power, long-life and high-repetition-frequency switching technology. Therefore, the development of various types of switching elements with resistance to high voltage and strong current, short breakdown delay, small dispersion, small inductance and resistance, less electrode burnout, and stable operation under repeated pulses is another important research topic in pulsed power technology abroad and internationally.
4. Actively open up new applications. As mentioned earlier, pulsed power technology has been increasingly used in the fields of nuclear physics, accelerators, lasers, electromagnetic emissions, and so on. In recent years, the research of the application of pulsed power technology in the fields of semiconductors, integrated circuits, chemical engineering, environmental engineering, and medical care has attracted wide attention from all walks of life, and some progress has been made. With successful application experience, pulsed power technology will be more applied to civil technology. The civilian market has great potential, and the promotion of the market will definitely bring new vitality to the development of pulsed power technology.

Chapter 2
Basic Theories of Pulsed Alternators

Inertial energy storage is an energy storage technology that stores energy with moving objects. It is often used in homopolar or pulsed alternators. This book didn't mention homopolar alternators due to their low output voltage, while focused on pulsed alternators instead. Pulsed alternators usually take rotors as their energy storage body under the speed of thousands of revolutions or even tens of thousands of revolutions per minute, allowing for the storage of large amounts of energy. There are many advantages of inertial energy storage technology, such as high energy storage density, compact structure, small size, low cost and high mobility. It has been applied wildly in the fields of industry, military and aerospace.

2.1 Principles of Pulsed Alternators

The main disadvantage of a homopolar alternator is its fairly low voltage. Although the synchronous generator can produce a higher voltage, an increase is required in the magnetic field in field coils or the speed or the number of armature winding turns. The first two factors are limited by the physical properties of the material used, and more turns will lead to the squared times of the inductance the winding coils, which will cause longer risetime of the synchronous generator voltage.

The University of Texas at Austin-Center for Electromechanics (UT-CEM) W. F. Weldon et al. invented the Compensated Pulsed Alternator (short for CPA) based on alternator, which overcame many of these shortcomings. They connected a fixed coil which was almost identical to the rotating coil in series and made the two coils coaxial. When the rotating coil coincides with the surfaces of the two coils, the inductance of the rotating coil is compensated, resulting in a strong electrical pulse. After the pulse peak, the inductor increases for the second time to its initial value as the coil rotates. The principles of electromagnetic induction and flux compression are combined in CPA for the integration of inertial energy storage, electromechanical

S. Wu and S. Cui, *Pulsed Alternators Technologies and Application*,
https://doi.org/10.1007/978-981-33-4224-8_2

energy conversion and pulse forming. It boasts an output of tens of kilovolts of high voltage and a faster rise time without the requirement of a switch with a large current capacity. In this way it can automatically change direction and produce repeated pulses. In addition, its slender rotor makes it easy to achieve high-speed rotation. In the continuous pulsed operation, because of its large inertia of rotation, there is no significant shrink in speed, therefore the next pulsed operation will not be affected. Flywheels are not usually used to balance the unit, and the hinge only plays a supporting role and does not withstand a large impulsive torque. CPA also has the advantages of simple structure, small size, light weight, convenience in operation and reliability in operation.

2.1.1 The Basic Principles of Conventional Generators

Pulsed Alternator (PA) is structured like conventional generators, containing stators and rotors, and the voltage is also produced in the same way as conventional generators, according to the electromagnetic induction principle. First of all, we should introduce the basic principles of conventional generators.

A generator is a device that converts mechanical energy into electrical energy, based on the electromagnetic induction principle discovered by Faraday in 1831. There is a change in the flux through the coil that senses the power generation in the coil. Figure 2.1 shows a schematic of a simple generator, due to the rotation of the

Fig. 2.1 Schematic diagram of conventional generators

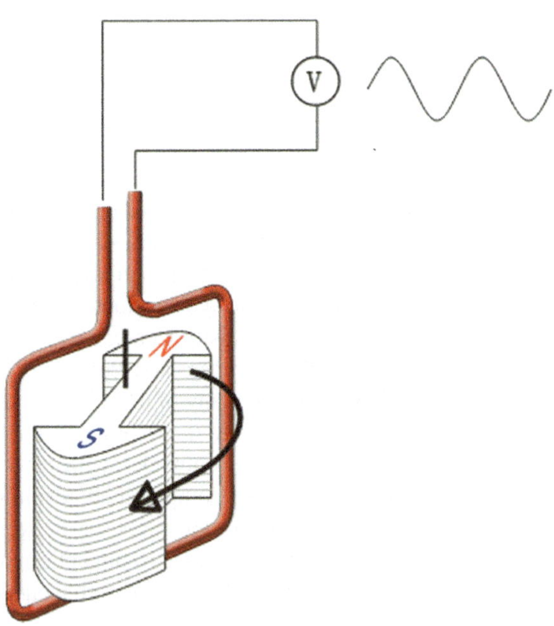

permanent magnet, the flux through the coil changes, the coil therefore senses the current. The earliest generator, invented by Faraday, is called the Faraday disk, which consists of a copper disk and a pair of permanent magnets, where the copper disk rotates in the middle of the vertical and horizontal permanent magnet, generating DC voltage. The first rotary motor was made by Agoby in 1834. The first British motor was recorded in 1856, but it wasn't put into practical use until 1873 in Westminster. The first electric light was invented in 1881, whose electric power was generated by the water wheel-dragging motor. After that, electricity was quickly used all over the world, and a variety of generators were invented and manufactured.

Depending on the form of voltage or current output, the generator can be divided into DC generators and alternating generators. According to the relationship between motor speed and synchronous speed, the alternator can be divided into induction generator and synchronous generator. Synchronous generators, driven by electric motors and other machinery, including internal combustion engines, water turbines, steam turbines, etc., when they are in a steady operation, the speed of their rotors is the same as the synchronous speed. It is widely used in power grids, automobiles, ships and other electrical equipments.

In terms of structure form, the synchronous generator can be divided into rotating armature generator and rotating magnetic pole generator. For the former one, its armature is installed on the rotor, and its main magnetic pole is on the stator. It's used more often in small synchronous generators. While in high-voltage large synchronous generators, the latter one is preferred. Since the capacity and voltage of the excitation part is much smaller than the armature, the load of the armature and the collecting ring can be greatly reduced and the operation can be more reliable and stable by installing the main magnetic pole on the rotor.

As for the rotating magnetic pole structure, according to the shape of the main pole, it can be further divided into non-salient and salient rotor. The non-salient rotor is cylindrical and with air uniform. The salient rotor has a clear protruding magnetic pole, and the air gaps are not equivalent. For high-speed synchronous generators, the non-salient rotor is used in terms of the mechanical strength of the rotor and the fixed excitation windings.

2.1.2 The Basic Structure of the Pulsed Alternators

The difference between pulsed alternators and conventional motors mainly lies in the application of special compensation components, so the pre-study of pulsed alternator mainly refers to the compensated pulsed alternator. Figure 2.2 shows a structural diagram of the iron-core compensated pulsed alternator. Iron-core machine refers to the motor whose yokes of internal stators and rotors are made of the ferromagnetic material with the high magnetic conductivity. Since the ferromagnetic material has a high magnetic conductivity, the excitation magnetic field required is not very small,

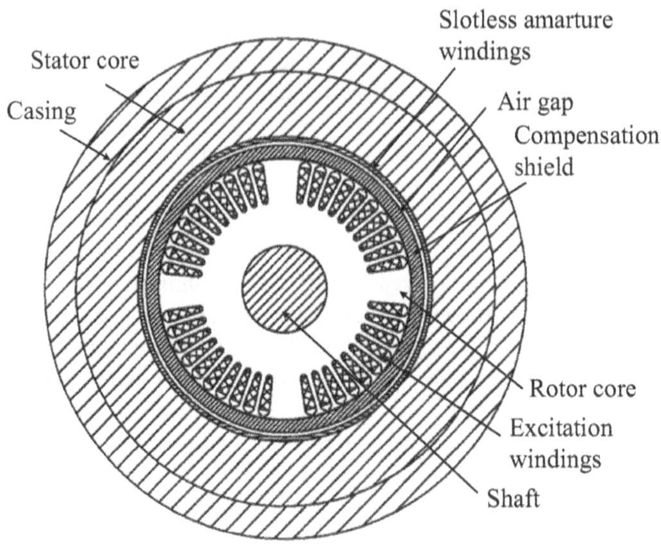

Fig. 2.2 Schematic diagram of the structure of the iron-core compensated pulsed alternator

no self-excitation magnetic with higher standards of control is required. The disadvantage is from the large proportion of ferromagnetic material, and the low motor's energy density and power density.

Pulsed power supply requires a high energy density and power density to fulfill the needs of the load. In this case the stators and rotors of the pulsed power alternator is made of non-magnetic materials (such as fiber resin composite materials or non-magnetic alloys, etc.) instead of the traditional ferromagnetic material for motors. The motor made is called the air-core compensated pulsed alternator. Its structure is shown in Fig. 2.3. On the one hand, the composite is of small density and great intensity. The motor can run with a line speed more than 500 m/s, far more than the limit speed of ferromagnetic material (125 m/s), and the energy storage density is significantly increased; on the other hand, the composite is unconducted magnetic, therefore the motor is not limited from magnetic saturation. The air gap flux density can be designed to extremely high (even 4-5T) to obtain higher power density. In iron and air-core machines, the armature windings are in the form of slotless windings, which can reduce the inductance of the windings.

2.1.3 Working Principle of Pulsed Alternators

The internal impedance of conventional generators is large. When discharging to loads with low impedance such as electromagnetic guns and electric cannons, it is difficult to obtain the pulses of narrow width and high amplitude required for the load. W. F. Weldon et al. from The University of Texas Center for Electromechanics

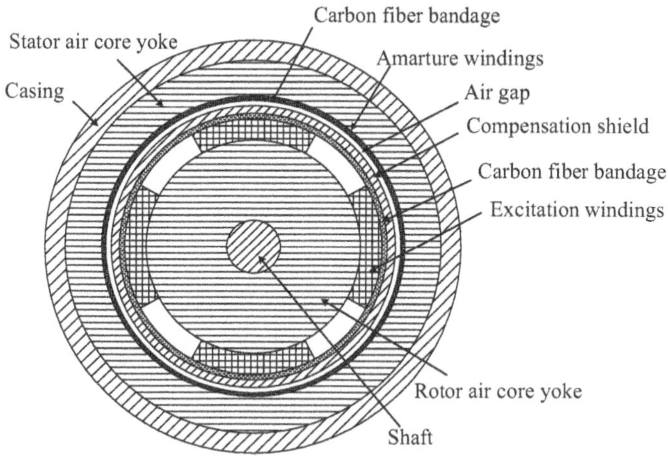

Fig. 2.3 Diagram of the structure of the air-core compensated pulsed alternator

invented compulsators in 1978, using the flux compression principle to maximize output current, and instantaneous output power and minimize the internal impedance. A compensating current is generated within the compensation component during the operation of the load, which produces a magnetic field that interacts with the armature reaction magnetic field, therefore reducing the instantaneous inductance of the pulsed alternator armature winding, while compressing the magnetic field to obtain a higher end voltage. This leads to a strong current pulse with extremely high peaks for driving the load. To know more about the basic method of analysis, readers can refer to the law of magnetic chain conservation of superconducting circuit, simplified circuit model and field analysis as follows.

1. Analysis of the law of pulsed alternators with magnetic chain conservation law. As can be seen from the laws of the circuits, any self-closing coil with a chain flux can be applied to the following equation:

$$ri + \frac{d\Psi}{dt} = 0 \tag{2.1}$$

In this equation, the magnetic chain of the closed coil includes the self-chain and the inter-chain with r for resistance and i for current.

If the resistance r is omitted, it can be derived from the upper formula that Ψ is a constant. It can be seen that the magnetic chain remains unchanged in a closed circuit (also known as a superconducting circuit) without resistance. If the external flux enters the coil, there will inevitably appear a current immediately, which produces a flux of same value in the opposite direction to keep the total flux of the coil linkage unchanged. This is the conservation law of the flux linkage of the superconducting closed circuit. In the actually closed circuit, the flux linkage changes due to the influence of the resistance. However, in the first period, the flux linkage of the superconducting circuit magnetic chain remains conservative. Therefore, taken the conservation of the flux linkage in consideration, the basic method of analyzing a sudden short circuit is to firstly calculate the current at the moment of the sudden short circuit and then to put role of the resistance to take into account. Under the affection of winding resistors, the current that occurs during transients will eventually be attenuated to a steady state short-circuit current.

The status of the normal synchronous generator at the short-circuit moment is shown in Fig. 2.4, and its equivalent circuit is shown in Fig. 2.5.

X''_d is ultra-transient reactance, X_d' is transient reactance, and X_d is direct-axis reactance. Apparently, $X''d < X_d' < X_d$. The compensation shield of the compensated pulsed alternator is equivalent to a very strong damping winding, and when there appears a short circuit, it is the eddy current that compresses the armature reaction flux into the gap between the compensation shield and the armature winding, thus reducing the armature winding.

2. Simplified circuit model analysis of the principle of pulsed alternators

CPA is generally simplified into three coupling loops in terms of the circuit: excitation circuit, armature circuit, and compensation circuit, where excitation circuit

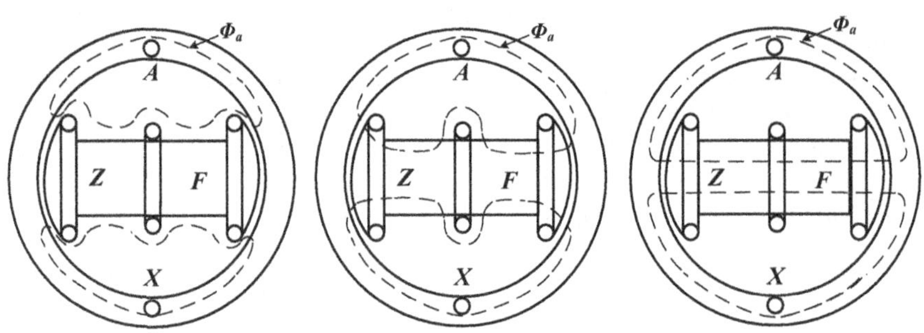

Fig. 2.4 A diagram of part of the magnetic field of a sudden short circuit of the synchronous generator

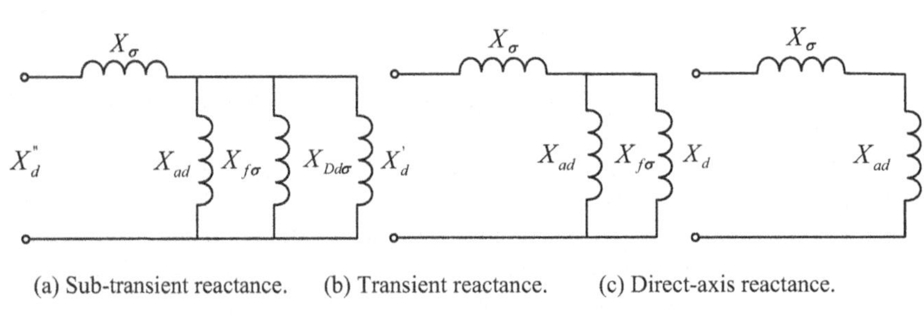

Fig. 2.5 The equivalent the ultra-transient, transient, and direct-axis reactance of a sudden short circuit of the synchronous generator

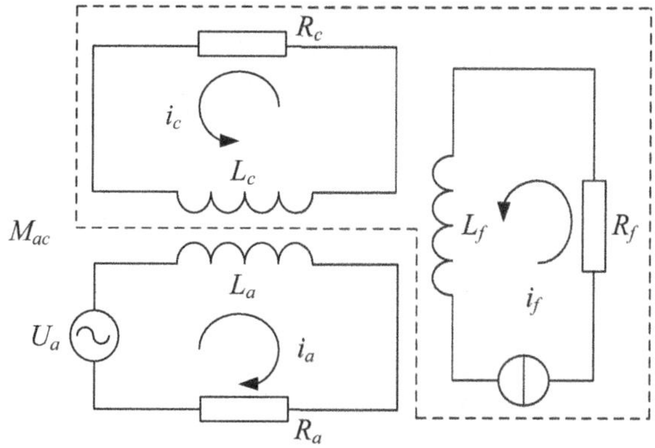

Fig. 2.6 The simplified model of CPA

and compensation circuit are always on one side of the motor. The simplified circuit model can be shown in Fig. 2.6.

In external constant current source excitation, the excitation current at the moment of discharging (i_f) remains unchanged, so the effect of the excitation circuit on the armature circuit and compensation circuit can be ignored; meanwhile, the internal reactance of CPA tends to be way smaller than the inductive reactance, therefore it is reasonable to be neglected. In this case, the voltage equation of the armature circuit and the compensation circuit can be expressed as:

$$\begin{cases} U_a = L_a\frac{di_a}{dt} + M_{ac}\frac{di_c}{dt} \\ 0 = M_{ac}\frac{di_a}{dt} + L_c\frac{di_c}{dt} \end{cases} \tag{2.2}$$

where U_a is the CPA output voltage; i_a is the armature current; i_c is the compensation current; L_a is the self-inductance of the armature winding; L_c is the self-inductance of the compensation winding; M_{ac} is the interaction between the armature winding and the compensation winding.

The formula (2.2) can be arranged in the following form

$$U_a = (L_a - \frac{M_{ac}^2}{L_c})\frac{di_a}{dt} = (1 - k_{ac}^2)L_a\frac{di_a}{dt} = L_{aef}\frac{di_a}{dt} \tag{2.3}$$

L_{aef}—equivalent inductance of the armature winding;
k_{ac}—coupling coefficient,

$$k_{ac} = \frac{M_{ac}}{\sqrt{L_a L_c}} \tag{2.4}$$

It's used to describe the compensation effect of CPA and $0 < k_{ac} < 1$.

After compensation, the equivalent inductance of the armature winding becomes ($1\text{-}k_{ac}^2$) times of the value before compensation, so at the same output voltage, a larger output current can be obtained, and the more powerful the coupling between the compensation winding and the armature winding is, the smaller the equivalent inductance and the greater the peak of the output pulsed current will become.

3. Field analysis of the principle of pulsed alternators

Field analysis of the principle of pulsed alternators contributes to a further research of its essence as an electromagnetic field. The electromagnetic field in the pulsed alternators changes instantaneously with time and belongs to the time-varying electromagnetic field but of low frequency. The displacement current density inside is much smaller than the conduction current density, so it can be neglectable, which means the volatility of the internal electromagnetic field can be ignored. The internal magnetic field of the pulsed alternators can be regarded as a static field, which is distributed similarly to the static field, only taking electromagnetic induction into account.

The operation of the pulsed alternators follows the conservation law of total flux. Under a no-load condition, the magnetic field in the compensation shield remains conservative, so there is no induction eddy current inside the compensation shield. At this time the addition of the compensation shield can be seen as an equivalent air gap that increases the size of the motor. During discharging, there is an alternate discharge current in the armature winding which is able to sense the eddy current inside the compensation shield. The eddy current field generated by the eddy current will compensate the armature reaction magnetic field, so that the armature winding will obtain a very low inner inductance. According to the conservation law of total flux, a high-amplitude pulsed current will be loaded by the armature. The principle of total flux conservation is summarized as below. The variable inductor $L(t)$ and the variable resistance $R(t)$ are circuited in series, whose current is $I(t)$. The Voltage equation of the Kirchhoff circuit is

$$\frac{d}{dt}[L(t)I(t)] + R(t)I(t) = 0 \tag{2.5}$$

When the variable resistance $R(t)$ is small enough to be ignored, the solution of the above equation is $L(t)I(t) = L(0)I(0)$, i.e. the total flux at any moment is equal to the total flux at the initial moment.

This compensation principle is similar to electromagnetic shielding. Electromagnetic shielding includes magnetic shielding and electromagnetic shielding. Magnetic shielding uses shielding material with high magnetic conductivity that shortens the magnetic field and reduces its influence on the external circuit. While electromagnetic shielding adopts the material with high conductivity which can sense the eddy current in the magnetic field that keeps changing, the magnetic field is blocked and the influence of the magnetic field is reduced on the external circuit. Magnetic shielding increases the inductance of the shielded conductor, while the electromagnetic shield reduces the inductance of the shielded conductor. Electromagnetic shielding doesn't

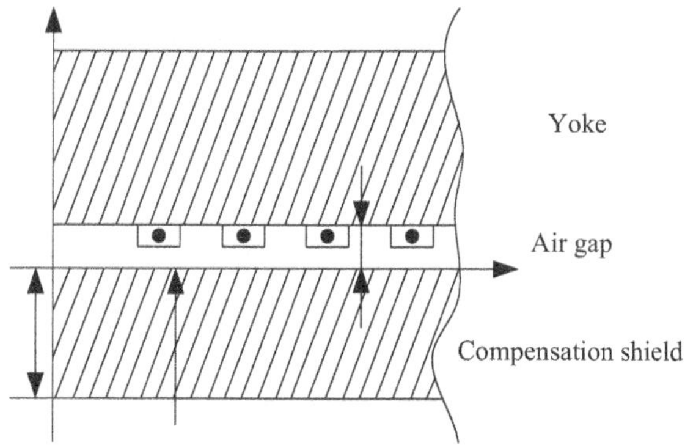

Fig. 2.7 CPA air gap and its sides expanded in the circumferential direction

impose an obvious effect on a low-frequency shielding but does the opposite to a high-frequency one. Therefore, it is applicable for pulsed power technology.

When designing pulsed alternators, the length of the air gap and the size of the armature winding are usually much smaller than the size of the internal diameter of the stator and the outer diameter of the rotor, so the air gap and its sides can be expanded in the circumferential direction, and the magnetic field in the motor air gap is seen as axially unchanged parallel plane field, as shown in Fig. 2.7. The air gap length is δ, the rotor's outer diameter is a, the radial thickness of the compensation shield is h, and the discharge current in the armature winding is i, wherein $\delta \ll a$.

It is supposed that the magnetic field strength of the discharge current i in the armature winding in the air gap is H_0, and the magnetic field strength at the inner wall of the compensation shield is H_c. According to the continuity conditions of the electromagnetic field, the difference in the strength of the magnetic field on both sides of the sub-interface is equal to the density of the current line on the sub-interface, i.e.:

$$H_c - H_0 = K_1 \tag{2.6}$$

K_1—the line density of the induced eddy current in the compensation shield;
E_1—the strength of the electric field in the compensation shield;
γ—the conductivity of the compensation shield;
h—the thickness of the compensation shield.

Taking the conductive circuit l_1 in the compensation shield into account, it can be obtained as follows according to Maxwell's equation

$$\oint_{l_1} \overline{E_1} \cdot d\bar{l} = -\frac{\partial}{\partial t} \iint_{S_1} \overline{B_1} \cdot d\overline{S} \tag{2.7}$$

Let (2.6) in (2.7),

$$\frac{K_1}{\gamma h} \cdot 2 \cdot \pi \cdot a = -\mu_0 \frac{\partial H_c}{\partial t} \cdot \pi \cdot a^2 \tag{2.8}$$

μ_0—permeability of vacuum.

Simplified (2.8) obtained as,

$$\tau \frac{\partial H_c}{\partial t} + H_c = H_0 \tag{2.9}$$

where $\tau = \frac{\mu_0 a \gamma h}{2}$. Based on solution of (2.9), the strength of the magnetic field in the compensation shield can be obtained

$$H_c = H_0 \left(1 - e^{-t/\tau}\right) \tag{2.10}$$

The line density of the induction eddy current in the compensation shield is:

$$K_1 = -H_0 e^{-t/\tau} \tag{2.11}$$

From the formulae (2.10) and (2.11), it can be seen that the induction eddy current inside the compensation shield decreases exponentially so that the internal magnetic field of the compensation shield will not mutate, hereby delays the diffusion from the armature reaction magnetic field to the inner compensation shield, and compresses the armature reaction magnetic field, in order to reduce the transient inductance of the armatures.

2.1.4 The Working Process of Pulsed Alternators

Unlike conventional synchronous generators, the basic steps for the operation of pulsed alternator systems includes energy storage, self-excitation, discharge, and energy recovery. Because the motor uses non-magnetic composite material, it is difficult to establish a magnetic field. The magnetic field is therefore established by self-excitation. Take a single-phase pulsed alternator for example. Its simplified circuit is shown in Fig. 2.8.

(1) Start-up process: the prime mover drives the pulsed alternator to the rated speed, and the rotors of the motor stores kinetic energy.
(2) Self-excitation process: close the excitation switch. A seed current with thousands of amperes is input to the rotor excitation winding by the pulsed capacitor, there comes the initial rotating magnetic field and induction current in the state armature winding. The armature current flows back to the excitation winding through the external self-excitation rectifier. Under certain conditions where the circuit parameters meet, the excitation current gradually increases, forming a positive feedback self-excitation process, and the mechanical energy

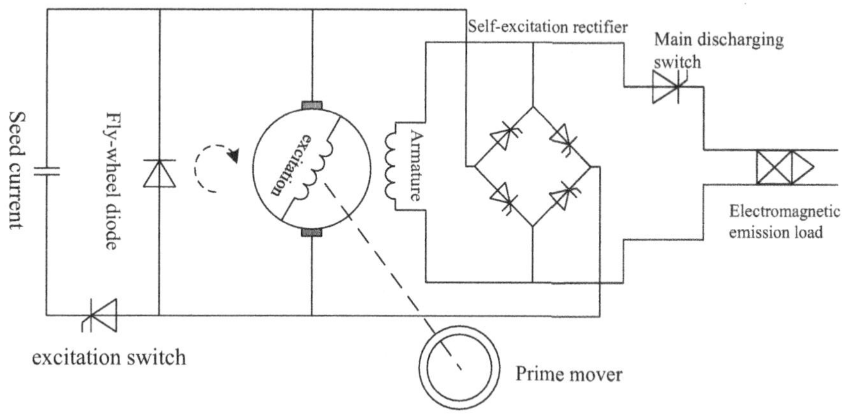

Fig. 2.8 The diagram of the circuit of a single-phase pulsed alternator

of rotor storage is converted into the energy storage of magnetic field. Since the pulsed alternator is not affected by magnetic saturation, the excitation current can theoretically increase exponentially.

(3) Discharging process: when the rated excitation current is reached, the trigger signal of the self-excitation rectifier will be stopped, and the excitation winding continues to provide the rotating excitation magnetic field by the reflux of short diode. The main discharging switch is triggered in the appropriate phase according to the load requirement, and the armature winding discharges to the load.

(4) Energy recovery: after discharging, there is still current in the excitation winding in the continuous flow, this part of energy storage in the magnetic field can be released by the reflux or external discharge resistance. After the excitation current is reduced to zero, a discharge ends, and the motor rotates freely by inertial energy storage, or replenishes energy by the prime mover, waiting for the next firing command.

A typical excitation winding current is shown in Fig. 2.9.

2.2 Type of Pulsed Alternators

According to the different materials used for the yokes of stators and rotors, it can be divided into iron-core motors and air-core motors. The iron-core motor refers to the motor whose material used in the yokes of stators and rotors is ferromagnetic material, while the air-core motor refers to motor whose material used in the yokes of stators and rotors is or partly is high-strength non-magnetic material. The motor with air-core structure can improve the strength of the yokes of the stators and rotors,

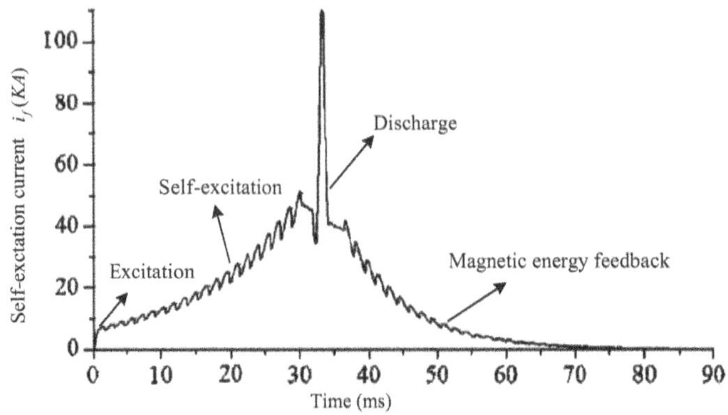

Fig. 2.9 Typical excitation winding current waveform

so under the same structure, the motor can be accelerated, and the energy storage of the motor can be enlarged, and the quality of the motor is reduced to improve the energy storage density.

The compensation element of the pulsed alternators is of a variety of structural forms and connection methods, which drive different relationships between the change of inductance and the rotor rotations, thus outputs different pulsed current waveforms. The three categories of compensation are: passive, active and selective passive.

2.2.1 *Classification of Compensation*

1. Passive compensation

The compensation shield (aluminum shield in the most cases) with uniform thickness and good conductivity provides uniform compensation for the armature winding, as shown in Fig. 2.10. The compensation shield is mounted between the excitation winding and the armature winding and remains relatively static with the excitation winding. Since the compensation shield is continuous, the coupling coefficient k_{ac}

Fig. 2.10 Diagram of the passive compensation structure

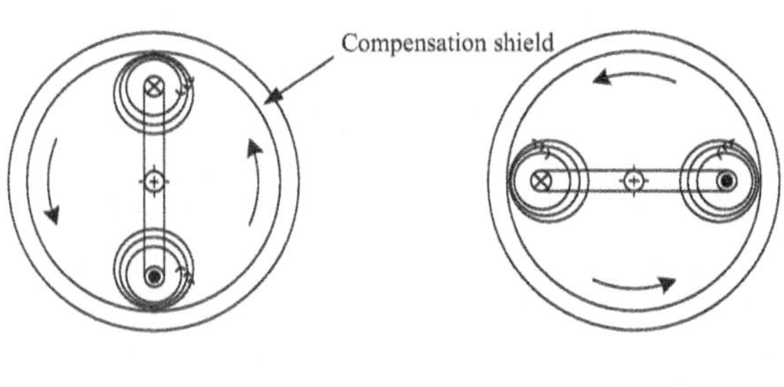

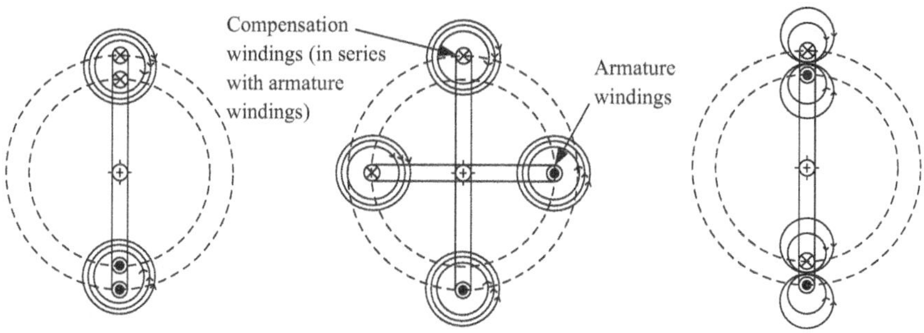

Fig. 2.11 Diagram of the active compensation structure

always maintains a large value regardless of the rotor's position, which gives the armature winding a constantly low inductance, and approximates the output current to be a sine wave (suitable for electromagnetic emission).

2. Active compensation

A set of compensation windings is dedicated and connected in series with the armature series by a slip ring and brush, as shown in Fig. 2.11. Under load, the mutual perception between the armature winding and the compensation winding changes periodically with the relative motion between stators and rotors. When the two winding axis is reversed, the coupling coefficient is the largest, at this time the inner inductance of the winding is approximately equal to the sum of the inductive leakage of the two sets of windings. This brings about a very narrow spike pulse which is suitable for flashing load and so on.

3. Selective passive compensation

The compensation winding is achieved by self-short-circuiting, as shown in Fig. 2.12. Because the compensation current is generated by induction, the magnetic field generated by the compensation winding always weakens the magnetic field of the

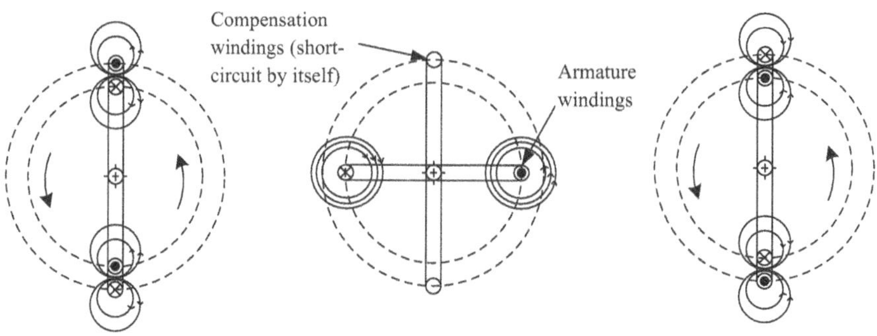

Fig. 2.12 Diagram of the selective passive compensation structure

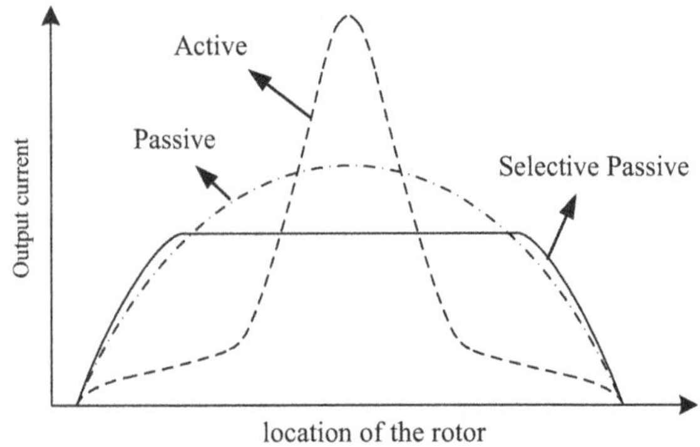

Fig. 2.13 Typical pulse current waveforms in three compensation methods

armature reaction. The coupling coefficient of armature windings and compensation windings, k_{ac}, is determined by the relative position between them. The internal inductance of the motor varies periodically with the position of the rotor. After special design, a rectangular-like current pulse can be obtained. This is exactly the ideal current waveform required for the load of the electromagnetic orbital gun.

The typical current pulsed waveforms in the three compensation methods are shown in Fig. 2.13.

4. No compensation

The compensation component reduces the inductance in the motor, but also increases the structural complexity. Pulsed alternators with no compensation do not require special compensation components. They use excitation windings as compensation elements instead and discharge through the armature reaction by compression of the armature winding flux, to reduce the inductance in the motor.

2.2.2 Classification of Excitation

In addition to the form of compensation, pulsed alternators can also be classified according to the excitation method, i.e. electrical excitation, permanent magnet excitation and hybrid excitation.

1. Electrical excitation

Electrical excitation refers to the use of excitation current to produce excitation magnetic field. For the iron-core machine external excitation or self-excitation is preferred, while for the air-core machine auto excitation or capacitive pulsed excitation is more suitable.

2. Permanent magnet excitation

Permanent magnet excitation refers to the use of permanent magnetism to produce an excitation magnetic field only with permanent magnets, as is the case with conventional permanent magnet motors. Compared with electrical excitation, the motors made of high remanence, high coercivity and rare earth with high magnetic energy product are advantageous because of their simple structures, reliable operations, small sizes, light weights, small losses and high efficiencies. However, because the magnetic characteristics of the permanent magnet motor are difficult to adjust, when the speed or load changes, the voltage of the alternator is difficult to maintain, hereby affects the practical uses.

3. Hybrid excitation

Hybrid excitation refers to the use of both permanent magnets and electrical excitation to establish the magnetic field inside the motor. A strong magnetic field to meet the requirement of motor discharge is established by permanent magnets and electrical excitation windings. The CPA achieves brushless, improves motor reliability, is more suitable for working at high speed, and increases energy density and power density. The electrical excitation load is reduced in this case, which improves the winding heat and the motor efficiency. By adjusting electrical excitation, the magnetic field can be adjusted quickly and flexibly, the motor outputting performance can be adjusted quickly, and different kinds of driving pulse current waveforms suitable for different types of pulsed loads can be obtained, so as to fulfill multi-purposes.

2.3 Development of Pulsed Alternators

Compensated pulsed alternator is a pulsed power supply which integrates inertial energy storage, electromechanical energy conversion and pulse forming, and has comprehensive advantages in the aspects of power density, energy storage density, high repetition frequency and service life compared with other pulsed power supply (i.e. inductors, capacitors, homopolar generators, etc.). It is more and more popular in the military and civil field. With the development of compensated pulsed alternator and to be more combat-oriented, the design of compensated pulsed alternator targets at high-speed, lightweight, integration, modularization and generalization.

Currently, major research institutions around the world include Center for Electromechancis at The University of Texas at Austin, Lawrence Livermore Laboratory, Culham Laboratory in the United Kingdom, Loughborough University of Technology, The All-Russian Experimental Physical Science Research Center, the Electrical Research Center of the Russian Academy of Sciences, the Huazhong University of Science and Technology and the Institute of Plasma Physics of the Chinese Academy of Sciences. The United States carried out the biggest investment and reached the highest level of research among them.

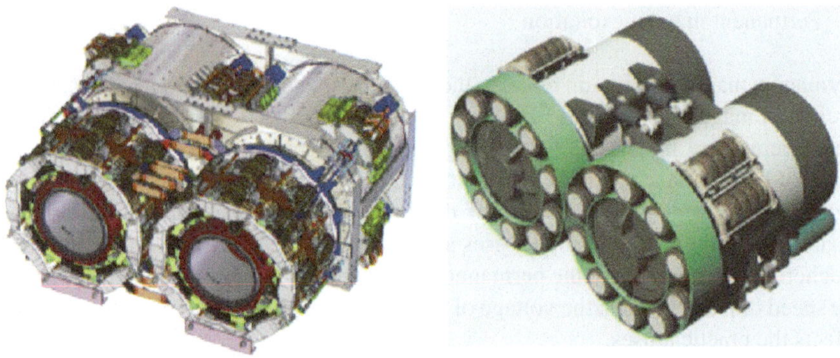

Fig. 2.14 A 3D diagram of the CPA set

The CEM of University of Texas undertook the Key Technology Program for Pulsed Power Research (Term 6.2) in the U.S. Army Electrical Weapons Program, which aims to develop the next generation of high-power, high-energy-storage density pulsed alternators. Under the U.S. military plan, the electric gun system, powered by a new generation of pulsed alternators, will be equipped with "the main battle tanks in the future." Figure 2.14 shows a 3D map of the pulsed alternator set designed by the CEM of University of Texas.

They have developed five generations of prototypes:

1. First engineering CPA prototype

In 1978, UT-CEM designed the first engineering CPA prototype, as shown in Fig. 2.15. The original design was designed to provide the appropriate high-power pulsed power supply for xenon lamps of the inertial-constrained nuclear fusion laser unit at Lawrence Livermore National Laboratory (LLNL) in the United States, and to solve the problem that the capacitive power supply does not provide the repetitive pulses required for xenon lamps. CPA can provide a certain amount of repetitive pulses with widths ranged from 500 s to 1 ms required by xenon lamps. The prototype is an iron-core vertical structure with single-phase four-pole, and a maximum speed of 5400 rpm, a no-load voltage of 5.7 kV, a storage energy at designed speed of 3.4 MJ, a peak value of discharge current of 72kA, and a pulse width of 560 μs.

The prototype rotor is stacked with No. 2913 silicon steel sheet to reduce the eddy current, and the heat sleeve is mounted on a hinge made of No. 4340 high-strength steel. In order to prevent high-speed rotation, the rotor lamination string affects the stability of the rotation of the rotor of the motor, using mechanical means to improve the hardness of the rotor, to prevent the rotor lamination separate from each other at 5400 rpm due to high-speed centrifugal force. The design of the yoke of the stator is capable of withstanding a discharge current whose peak value is150 kA. The pulsed torque on it is 2.7×10^6 N·m and the internal pressure is 20.7 MPa. Through the mechanical structure the stator yoke will be strengthened to prevent the destruction of the epoxy resin bonding layer between the stator yoke and excitation

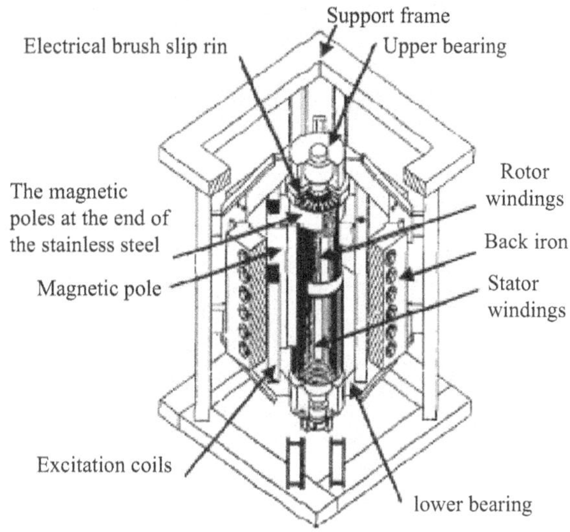

Fig. 2.15 First CPA engineering prototype

windings out of the sliding of the yoke. The outer frame of the stator yoke can be rotated, and when discharging the outer frame rotates synchronously with the yoke, and the torque will be reduced from 6.82×10^6 N·m to 1.15×10^5 N·m.

The most successful discharge experiment was under an acceleration to 4800 rpm, a peak current of 30 kA was emitted, with a pulse width of 1.3 ms, and a single pulse output energy of 140 kJ. The prototype successfully accelerated to a maximum speed of 5400 rpm under the absence of excitation, and when full of excitation, the voltage peak reached the design value of 6 kV. However, before getting ready to discharge, the motor was damaged due to the failure of the insulation at the end of the armature winding. Due to the eddy current loss produced by the solid pole of the stator, the compression ratio of the inductance is reduced, the CPA is structurally improved, the compensation winding is replaced with a compensation shield, and the armature winding is connected in series, while changing the size of the armature conductor and improving the insulation system of the winding in order to create a new prototype.

The CPA successfully verified the principle of active compensation, reduced the internal impedance of the motor to a very low value by using compensation windings, and simultaneously mediated the shape of pulsed current waveforms. It verified the accuracy of the program of winding resistor inductance and circuit simulation program prepared by UT-CEM and LLNL with spatial harmonic field distribution.

2. Rapid fire CPA

Pulsed alternators, as a "single element" which emits high-energy and high-power pulses based on the principle of compensation (magnetic flux compression), were originally used as a driving power for xenon lamps as inertial-constrained nuclear fusion laser devices, but after that were successfully applied by the U.S. Army and NASA into driving electromagnetic catapults and high-frequency microwave

Fig. 2.16 Rapid fire CPA

weapons. With the "Star Wars" program proposed by the United States in the 1980s, the U.S. Army established the Electric Artillery Project and began the development of a CPA for rail gun power, which was called iron-core compensated pulsed alternator (Iron-Core CPA, ICC) and was also known as the Rapid Fire CPA. It is the first CPA to be used directly to drive electromagnetic rail guns. The relative design was completed in 1983 and UT-CEM carried out the prototype discharge experiment in 1987. As shown in Fig. 2.16, the use of 30 mm dual-tube rail gun and 80 g mixed solid armature at a launch speed of 2 km/s firstly verified the feasibility of CPA as a rail gun power supply. The CPA is an iron-core transition with a single-phase, six-pole, and passive compensating structure. The rotor adopts solid AISI 4340 steel, and the stator is stacked with silicon steel sheet to reduce the eddy current. The armature windings are fixed to the inner surface of the stator yoke with fiberglass epoxy resin. 7050-T74 aluminum compensation shield is connected to the rotor by shrink fit. In this process, the match surface is coated with epoxy resin to improve the tightness of the fit and to prevent the aluminum shield and the rotor from getting loose because of the high-speed rotation. The most challenging part during design and processing is about the compensation shield. There are some requirements for the thickness of the material. It needs to meet the requirement of conductive compensation shield and compensate the flux to obtain the desired inductance. It also needs to consider its own strength to prevent high-speed cracking deformation.

The CPA pulse width is about 2 to 3 ms, single-shot emission energy is about 160 kJ, and the transmission power is 1.2 GW. This is the highest power that the existing synchronous generators can ever have. However, its energy density and power density still do not meet the standards of the U.S. On September of 1986, the CPA's initial full-speed experiment, due to the epoxy failure of the end area of the excitation coil, the excitation coil deviated to the compensation shield, so that part of the compensation shield was biased towards the armature winding, and finally

Fig. 2.17 Small caliber CPA-SCC

contacted with the armature winding. This leads to a destruction of motor winding insulation, excitation windings, armature windings and compensation shields. After a reliable launch of the rail gun at a frequency of 10 Hz, the ICC is used as an experimental power supply for high-power switches and rail gun tubes for new pulsed alternators.

3. Small caliber CPA-SCC

After the ICC successfully drove the electromagnetic rail gun and verified its working principle of compensation, the U.S. Army began to develop a more compact compensatory pulsed alternator with less weight. In 1988, UT-CEM designed and built the first air-core CPA, known as the Small-caliber CPA, as shown in Fig. 2.17. It used the composite material of high intensity-density contrast instead of the ferromagnetic material for ICC. Although it was rated at only half the power level of the latter and its mass is only 8% of which, the power density of SCC was 6 times higher in comparison. These confirm that the introduction of new composite materials significantly increases the energy density and power density of the motors. Afterwards all the generators studied by UT-CEM are air-core machines.

The researchers carried on the detailed analysis and design of Small Caliber CPA-SCC before processing and optimized the rotor with the programming. The design of CPA incorporates the most advanced technologies of the time, including a multi-layer composite rotor. Silicon nitride ceramic shafts and mixed silicon nitride bearings were adapted. The case is made of titanium.

Another feature of the small caliber CPA-SCC is its multi-layer composite rotor. The rotor consists of 15 layers, two of them are the armature winding layers, two of them are the layers of glass epoxy resin cloth, two of them are the fiber layers of axial fiber. The remaining 9 are the layers of the circular winding fiber winding.

The 9-layer circular winding layers are under interference fit and the excess is accurately calculated to ensure that each layer maintains radial pressure at high speed. The small-caliber CPA winding topology is characterized by two sets of armature windings on the rotor, one for self-excitation magnetism, and one for discharge to the electromagnetic gun. The two windings are designed according to their different operating currents and working hours. This realizes the optimal design of the motor and improves the excitation efficiency while ensuring the charging ability of the motor to the load remain unchanged.

4. Cannon caliber CPA-CCEMG

In the early 1990s, the U.S. Navy, after the U.S. Army, also developed a strong interest in electromagnetic guns, jointly set up a project to develop the Cannon-caliber electromagnetic gun system. The electromagnetic gun system index is set according to the actual military needs. At the same time, it requires that the size and weight of the system can match with amphibious assault vehicles. Under such demand, UT-CEM developed the Cannon-caliber CPA-CCEMG, as shown in Fig. 2.18. The single-store inertia of the CPA rotor can support 5 automatic fires of integrated launch projectile of 185 g at 5 Hz frequency from a rectangular electromagnetic gun (equivalent to 30 mm round gun). There is a total of 3 rounds, with inter-wheel interval of 2.5 s and the gun speed of 1.85 km/s.

As a single-phase air-core structure, the Cannon-caliber CPA-CCEMG uses self-excitation. Single-phase armature windings deliver both excitation currents for excitation windings and discharge currents for electromagnetic guns. The dynamic strains and support systems of the rotors are studied in detail, and the torque velocity curve is actively controlled in the experiment, so that the rotor sways smoothly through the first critical speed, preventing the composite rotor from sharp vibration in an

Fig. 2.18 Cannon-caliber CPA-CCEMG

Fig. 2.19 Model scale CPA and its discharging system

approximately critical speed, and resulting in the failure of the rotor with composite material.

5. Model scale CPA

In the mid-1990s, the U.S. Army launched the "Focused Technology" research program of pulsed powers for the development of the next generation. They tried to make more compact pulsed powers with lighter compensation, which would enable the electric gun system to be compatible with the 2015 version of "future main battle tanks". As a part of the all-electric tank, the power supply can provide instantaneous high pulsed power for electromagnetic guns, electromagnetic armor, etc., as well as continuous power for locomotive acceleration, regenerative braking and electromagnetic suspension. UT-CEM undertook the project, designed the model scale CPA, and established a discharge system, as shown in Fig. 2.19.

The CPA is of an improved inner rotor topology structure of transition-type with four phases. Through the study of model scale CPA, the design and simulation program is developed to calculate the CPA performance accurately, which facilitates the development of the prototype of next generation.

6. Focused technology program CPA

UT-CEM's fifth generation CPA is structured with four phases four poles, and self-exciting non-compensated transition topology. The rotors are featured with hollow structure to strengthen the edge structure. Not much related literature can be found currently. The topology and parameters of CPA developed by UT-CEM for electromagnetic guns are as shown in Table 2.1.

7. Advanced focused technology program

The newer generation of CPA in the United States is transferred to enterprises for research and development. The prototypes in the advanced focused technology program are two motor sets with synchronously reverse rotation, as shown in Fig. 2.20.

Table 2.1 The specific parameters of each generation of CPA developed by UT-CEM

	CPA for rapid-fire guns	CPA for small-caliber guns	CPA for cannon guns	Model scale CPA
Number of phases	1	1	1	4
Number of poles	6	2	2	4
Excitation type	Separately excitation with transition	Self-excitation with rotation	Self-excitation with rotation	Self-excitation with transition
Compensation type	Passive compensation	Passive compensation	Selectively passive compensation	No compensation
CPA weight (kg)	11,000	750	2000	2046
Volume (m^3)	2.1	0.22	0.85	0.88
CPA speed (rpm)	4800	25,000	12,000	12,000
Air gap magnetic density (T)	1.6	2	2	3.5
Rotor Energy Storage (MJ)	38	250	13.5	40
Phase current peak value (kA)	944	3500	560	660
Phase voltage peak value (kV)	2	8	3.6	2.8
Emission energy (kJ)	160	9000	64	280
Peak power (MW)	1200	27,000	1500	2500
Power density (kW/kg)	109	2160	1580	1221.9
Energy density (kJ/kg)	3.4	20	13.5	19.5

The rotors and stators of the prototype are shown in Fig. 2.21 a, b, respectively. The main parameters of the prototype are as following: speed (12,000 rpm), energy storage (46 MJ), voltage (7 kV), single discharge current (800 kA), rotor length (1 m), and rotor rim speed (500 m/s).

In addition to the United States, other countries such as Britain, France, Germany, Russia, India and other military powers have also invested in the CPA focused technology research, but most of them concentrated on model verification, theoretical analysis and so on. China had witnessed studies on high-power pulsed power supply. In recent years, most of them focus on theoretical research, analytical calculation and the development of the principle prototype. The main research institutions for pulsed alternator research in China are the Institute of Plasma Physics of The Chinese Academy of Sciences, the Institute of Electrical Engineering of the Chinese Academy

Fig. 2.20 CPA prototypes in advanced focused technology program

(a) Rotor diagram (b) Stator diagram

Fig. 2.21 Rotor and stator of advanced focused technology program CPA

of Sciences, Huazhong University of Science and Technology, Harbin Institute of Technology etc.

In the early 1980s, China carried out a follow-up research on compensated pulsed alternators. The Institute of Plasma Physics of the Chinese Academy of Sciences has developed six various types of CPA model prototypes, designed and manufactured a 25 MW passive compensator CPA in the late 1980s, and successfully completed the continuous firing experiment of driving electromagnetic gun, with a radio frequency of 4 rounds per second to fire 10 projectiles out of the load at a speed of 2 km/s.

In the early 1990s, a 10 MW active compensation CPA was designed, and the concept of serial-stage operation, polarization and dual-machine shell of CPA was proposed as a successful drive for solid lasers. The Institute of Electrical Engineering

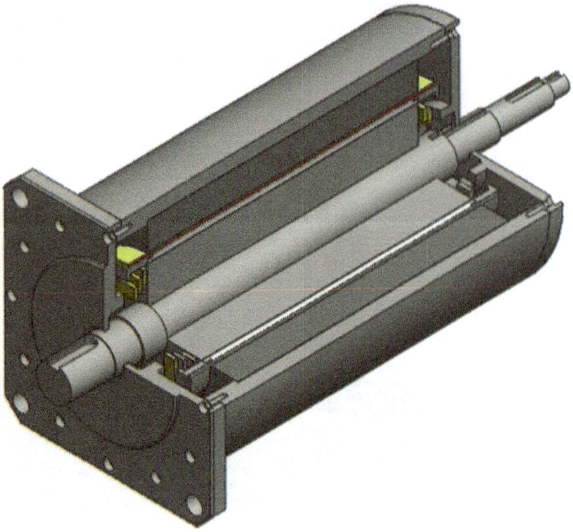

Fig. 2.22 Iron-core passive compensated pulsed alternator

of the Chinese Academy of Sciences designed an active compensated pulsed alternator with a compression ratio of 55 and carried out an experimental study on the capacity load. In the early 1990s, Huazhong University of Science and Technology started the research of CPA and developed two prototypes of active compensation and passive compensation respectively. Their research is mainly in theoretical analysis and experimental verification.

In China, Huazhong University of Science and Technology and Harbin Institute of Technology carried out the development of air-core prototype and yielded fruitful researches on the basic theory of air-core compensation alternators and engineering application. They put forward simulation calculation and experimental research on the motor's operating behavior under discharge, which laid the foundation for the theoretical research and engineering application of air-core compensated pulsed alternators.

The pictures of some of the prototypes that the studying team has developed are shown below. It mainly includes the iron-core passive compensated pulsed alternators (Fig. 2.22), the mixed excitation passive compensated pulsed alternator (Fig. 2.23) and the new air-core passive compensated pulsed alternator with double armature winding in stator (Fig. 2.24).

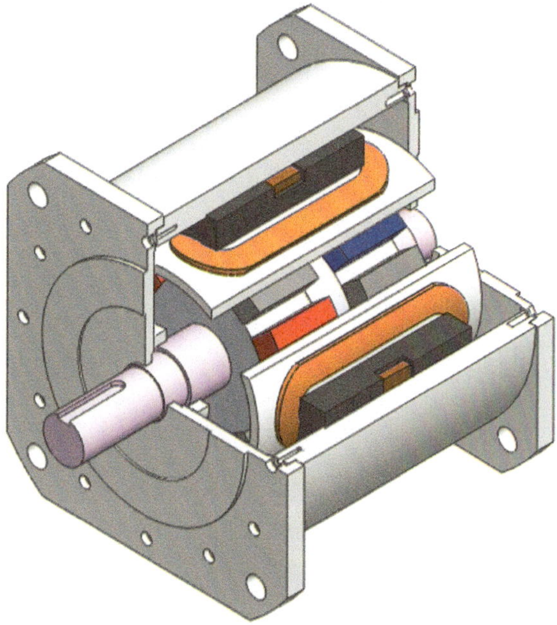

Fig. 2.23 Hybrid excitation passive compensated pulsed alternator

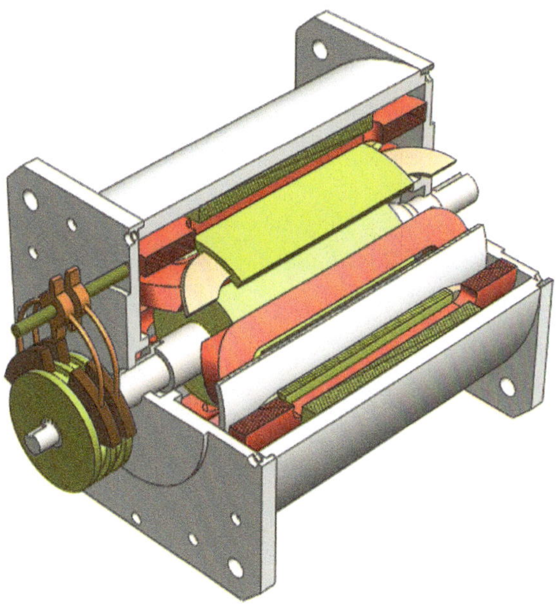

Fig. 2.24 Air-core passive compensated pulsed alternator with double armature winding in stator

Chapter 3
Electromagnetic Design of Pulsed Alternators

Because of their special structure and operating conditions, the pulsed alternators' design method and operating characteristics are different from the conventional ones. This chapter mainly expounds following aspects: the design methods of pulsed alternators, including the choice of poles, phases and main dimensions, calculation of electromagnetic parameters, analysis of no-load magnetic field and discharging characteristics, the establishment of mathematical model, the analysis of pulsed alternators by finite element modeling, and the introduction of its design process.

3.1 Relationship Among the Main Dimensions, Rotating Speed and Energy Storage and Power of Pulsed Alternators

The main dimensions of conventional motors refer to the specified stator inner diameters and armature length, which is calculated by the calculation formula based on the rating power of the motors. The design of pulsed alternators is the same as conventional ones. In the very process, the main dimension design needs to be carried out first.

The pulsed alternator sets inertial energy storage and electromechanical energy transformation as a whole, and its main dimensions need to meet the demand of energy storage and power. In some cases, the energy storage requirement is more demanding, therefore the priority needs to be given to the relationship between main dimensions and energy storage. While in others, it is more difficult to meet the requirements of power so the relationship between main dimensions and power is preferred for practical studies.

S. Wu and S. Cui, *Pulsed Alternators Technologies and Application*,
https://doi.org/10.1007/978-981-33-4224-8_3

3.1.1 Relationship Between Main Dimensions and Energy Storage

For pulsed alternators that need decent energy storage, the relationship between its main dimension and energy storage is prioritized. This type of pulsed alternators is typically used in the weapon system of a single drive with low energy but high emission frequency. It's difficult for the system to recharge the motor during the launch interval, so it is necessary to store enough energy in a single shot.

Pulsed alternators have two structures, solid and hollow, in order to improve energy density. They are generally designed as hollow structures, so the rotor can be simplified to a cylindrical structure. The rotor's inertia J is:

$$J = \frac{\pi}{2}\rho_r l_r (b_r^4 - a_r^4) = \frac{\pi}{2}\rho_r l_r (b_r^4 - b_r^4\lambda^4) \tag{3.1}$$

a_r the inner radius of the rotor;
b_r the outer radius of the rotor;
ρ_r the average mass density of the rotor;
l_r effective length of the rotor;
λ the ratio of the inner diameter to the outer diameter of the rotor.

Rotor energy storage E_r is calculated as:

$$E_r = \frac{1}{2}J\omega^2 \tag{3.2}$$

By formulae (3.1) and (3.2), the following rotor inertial energy storage calculation formula can be derived.

$$E_r = \frac{\pi}{4}\rho_r\beta\left(1 - \lambda^4\right)b_r^3 v_{tip}^2 \tag{3.3}$$

β the ratio of the long diameter of the rotor;
v_{tip} the line velocity of the rotor edge.

At this point, the rotor's energy storage density is

$$\frac{E_r}{m_r} = \frac{\pi}{4}\frac{\rho_r\beta(1 - \lambda^4)b_r^3 v_{tip}^2}{\pi\rho_r\beta(1 - \lambda^2)b_r^3} = \frac{1}{4}(1 + \lambda^2)v_{tip}^2 \tag{3.4}$$

By the formula (3.4), for a certain size of the rotor, the larger the internal and external diameter ratio is, (that is, the longer the inner radius is), the greater the energy storage density will be. According to the energy storage expression, the rotor energy storage decreases as the ratio of the inner and outer radius increases. Therefore, the inner and outer radius of the rotor should be properly selected. If the ratio of the inner diameter to the outer diameter ranges from 0.4 to 0.7, high energy storage density of rotors can be ensured while obtaining the required inertial energy storage. The long-diameter ratio is set in the range from 2 to 4 to prevent vibration at high speeds.

3.1.2 Relationship Between Main Dimensions and Power

The shear stress of the armature winding is a major limitation when the pulsed alternators need a high-power output. The pulsed alternator stator mainly consists of the armature winding and the stator yoke. The armature winding is bonded to the inner surface of the stator yoke, and the shear stress is calculated as follows

$$\tau_\theta = \frac{P_r}{\omega_f A b_r} \tag{3.5}$$

P_r the peak power of pulsed alternators;
ω_f the speed of the rotor after discharge, $\omega_f = \sqrt{1 - f_u}\omega_0$;
f_u The percentage of the inertial energy storage energy caused by discharges to the energy storage at full speed;
ω_0 the speed of the rotor before discharge;
A The effective bonding area of the winding.

The maximum shear stress that the winding can withstand is calculated by the utilization strength and safety coefficient of the bonding layer between the winding and the stator, and the effective adhesion area of the winding is calculated in turn.

When the pulsed alternator is used for long-range electromagnetic emission or electromagnetic weapons with the muzzle kinetic energy exceeding 10 MJ, the emission frequency is low and the single-shot energy is high, and the motor only needs to store the energy which is enough for a single shot. But in fact, because the peak power has a very high value, the armature winding is subject to a large shear stress. In order to reduce the shear stress, so that it will not exceed acceptability limit and the safety and reliability of operation of the motor can be ensured, we need to increase the effective adhesion area of the winding and the size of the motor. In this way, the energy stored by the motor will exceed the energy required for single-shot emission.

The speed of the motor is involved in the relationships between main dimensions of the motor and energy storage or power. Under the certain energy storage demand, the higher the motor speed is, the smaller the rotor radius will be. Similarly, under the certain power demand, the higher the motor speed is, the smaller the inner diameter of the stator will be. Therefore, the higher motor speed and the smaller the size of the motor is, the more beneficial it is to facilitate the miniaturization of high-power pulsed power supplies with inertial energy storage.

When the motor speed increases, the motor frequency will rise accordingly, resulting in an enlarged loss of the switching device, which requires more rectifier devices. In addition, increased frequency will also lead to an increase in the motor and load impedance. In order to meet the output current requirements, the need for higher output voltage is expected to be higher, and the motor design will be more difficult. Therefore, in the design and calculation of the main dimensions of the motor of the pulsed alternator, not only energy storage and power must be considered, but speed also needs to be considered comprehensively.

3.2 Principles for the Selection of Poles and Phases

After the main dimensions are determined, the pole and phase number of the motor also need to be considered. According to the parameters of the electromagnetic emission loaded device and the mass, acceleration and exit speed of the emitter, the pulsed current waveform required by the motor is calculated. The pole and phase number of the pulsed alternator are thus determined according to the parameters of the pulsed current waveform.

3.2.1 Pole Number Selection

For single-phase pulsed alternators, in order to obtain a wider pulsed current, the role pair number of the pulsed alternators should be as small as possible, however, there are many electromagnetic and mechanical problems in the actual operation of antipode motors. On the one hand, the rotor is unevenly affected. It is easy to cause polarization and stress concentration problems. On the other hand, in antipode motors' excitation windings, the internal magnetic fields are constant. They go through the motor shafts, as shown in Fig. 3.1, which are non-magnetic shafts and bearings in specific. Therefore, the single-phase pulsed alternator uses more than two pairs of polar structure. The flux path is short, and the winding is evenly distributed along the circumference. The electromagnetic and force distribution is symmetrical along the circumference.

With the widespread adoption of multiphase structures, the frequency of the motor is no longer limited by the pulse width, so under the frequency limit of the switching device, the pulsed alternator can theoretically adopt more polar designs to obtain the output current closer to the flat top ladder wave, but the more poles the motor has, the smaller the coupling coefficient between the armature winding and compensation windings will be, meaning the worse the compensation effect of the pulsed alternator will be. In this case, the multiphase pulsed alternator is usually designed as a two-pair or three-pair pole structure.

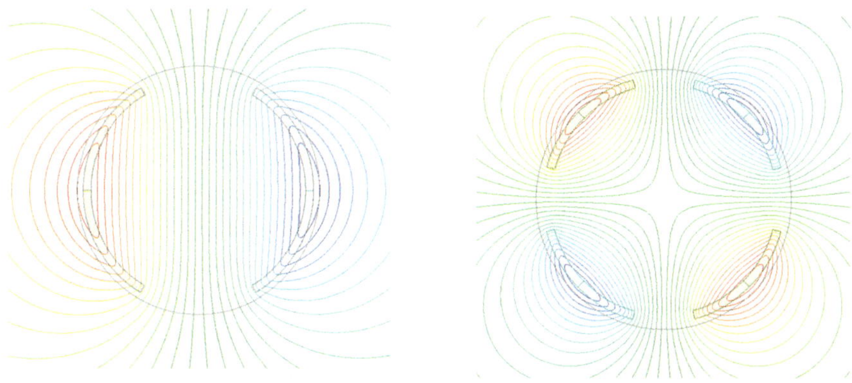

a) $p_0 = 1$ b) $p_0 = 2$

Fig. 3.1 Effect of pole pair number to the distribution of the magnetic lines of force in air-core pulsed alternators

3.2.2 Phase Number Selection

Pulsed alternators can be designed to be of single-phase or multiphase structures. The single-phase structure uses the positive half cycle of the output voltage to discharge to the load, and the output current consists of the single pulse itself. Because the discharge circuit is inductive, the pulse width of the output current is greater than half the electrical cycle. The single structure of the single-phase motor doesn't require much of the performance and control of the output switch device, and the electrical energy on the inductive transmitter such as the orbital gun can be automatically recovered after discharge. The disadvantage of the single-phase pulsed alternator structure is that the speed and frequency of the motor are limited by the pulse width, so it is only suitable for small-diameter rail guns with shorter pulse widths to drive, while it is difficult for large-diameter rail guns with long acceleration times to do so. At the same time, when a single-phase pulse alternator is discharged, the ignition angle is the only controllable degree of freedom, and there's no enough flexibility of the current waveform.

With a multiphase pulsed alternator, by controlling the closing angle of each phase winding, multiple short pulses will be synthesized into a wide pulse that fulfills load requirements, as shown in Fig. 3.2. This decouples the restrictive relation between the motor speed and pulse width. In this case, there will be a sufficient pulse width and an improved speed to obtain a higher energy storage density. At the same time, the output waveform will be more flexible, especially suitable for driving high-energy electromagnetic emission. In order to enable each phase to be independently controlled, two-phase and four-phase structures are preferred in multiphase pulse alternators. The electrical angle between each phase winding are quadrature so as to avoid the electromagnetic coupling between the phase windings.

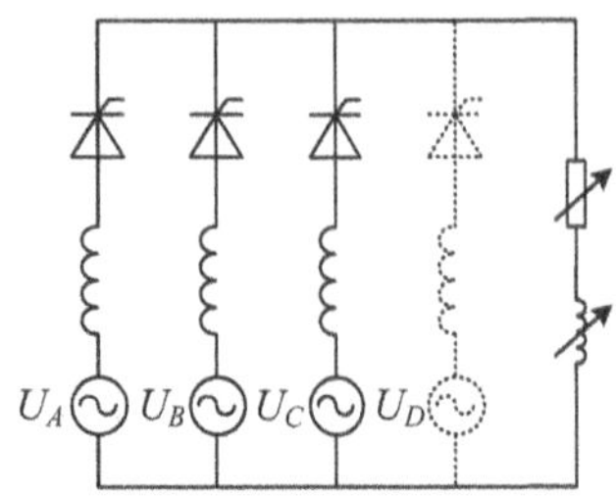

a) The circuit schematic of multi-phase pulsed alternators

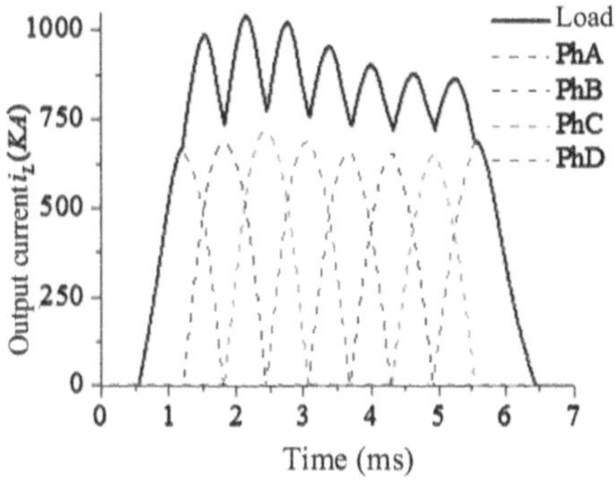

b) Output current waveform of typical four-phase pulsed alternators

Fig. 3.2 Multiphase structure schematic and typical output current waveforms

Different types of orbital guns require different types of launching. For a four-pole pulsed alternator, when the speed reaches 10,000 rpm, the output voltage of the electrical cycle is 3 ms, at this time the output pulse width of a single-phase motor range from 1.5 to 2.5 ms, just to meet the requirements of small-diameter guns and Cannon-caliber guns. A single-phase topology works better when driving an orbital gun which requires a short pulse width because there is no need to decouple the relationship between pulse width and speed. To drive the orbital gun with medium diameter or above, because the required pulse width time is long, only the multiphase structure can output the pulse current waveform to meet the demand for its load.

3.3 Analysis of the No-Load Magnetic Field of the Pulsed Alternators

When analyzing the traditional iron-core motor, the magnetic field is usually classified into the main magnetic field of the air gap and the leakage magnetic field, due to the small air gap and the magnetic nature of the ferromagnetic material. The stators and rotors of air-core pulsed alternators are both made of non-ferromagnetic material. The magnetic force line in the motor is not bound by the iron-core, therefore it is difficult to clearly distinguish the main magnetic field from the leakage magnetic field. The intensive storage of magnetic energy in the small air gap between the stator and rotor no longer exists, so the traditional iron-core motor analysis method must be replaced. In this case, it's necessary to study on the magnetic field distribution of the air-core pulsed alternators on the basis of electromagnetic field analysis.

3.3.1 Air-Core Alternator

Figure 3.3 shows a simplified model of pulsed alternator with pole number p. The relative permeability of the rotor yoke within the model is μ_r, and the relative permeability of the stator yoke is μ_s. To generalize the results, the radius of the winding is R_w, and neither the stator nor the rotor is in contact. When R_w equals to R_r, the winding represents the excitation winding; and when R_w equals to R_s, the winding represents the armature winding. In the air-core pulsed alternators, the relative permeability of the rotor yoke and the stator yoke are both 1.0.

It is also noteworthy to mention in the analysis that the influence of the end and the winding eddy current are both ignored. In order to simplify the calculation, the winding is replaced by a sine-distributed line current. The current is gathered around the centerline of the winding (ρ = R_w, ρ represents the polar radius on the polar coordinates). The line density of the current flowing through the winding is

$$K = K_m \sin(p\phi) = \frac{2Nk_w i}{\pi R_w} \sin(p\phi) \tag{3.6}$$

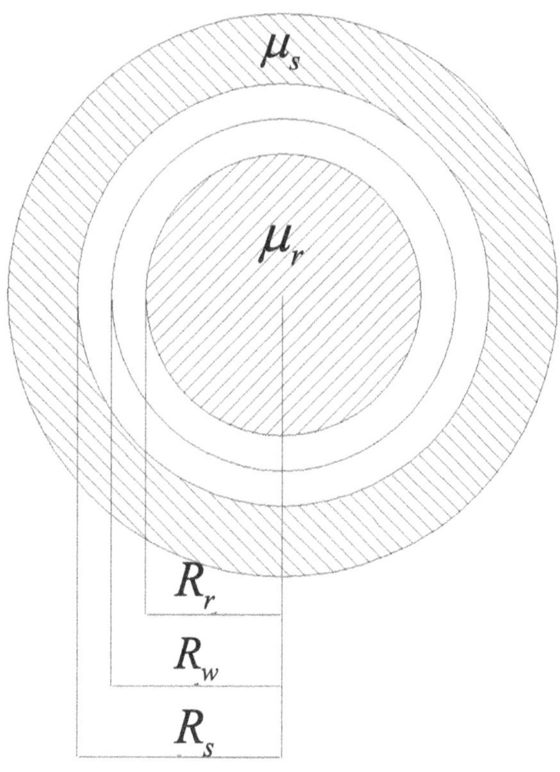

Fig. 3.3 Simplified model of air-core pulsed alternators

K current line density, unit is A/m;
ϕ mechanical angle;
N the total number of turns-in-series per phase winding;
k_w base wave winding factor;
i winding current.

Ignoring the edge effect, the magnetic field inside the motor can be regarded as a parallel plane field along the axis. Because the current i is in the axial direction, the axial component A_z is the only one on the magnetic vector potential, which meets the Laplace equation and can be expressed on the cylindrical coordinates as

$$A_z(\rho, \phi) = A_0(\rho)\sin(p\phi) \tag{3.7}$$

About $A_0(\rho)$, it can be expressed in the equation below,

$$\frac{d^2 A_0(\rho)}{d\rho^2} + \frac{1}{\rho}\frac{dA_0(\rho)}{d\rho} - \frac{p^2}{\rho^2}A_0(\rho) = 0 \tag{3.8}$$

Taking the boundary conditions into account,

$$\begin{cases} A_z(0) = 0,\ A_z(\infty) = 0 \\ A_z(R_{w+}) = A_z(R_{w-}) \\ \frac{1}{\mu_0}\frac{\partial A_z(R_{w+})}{\partial \rho} - \frac{1}{\mu_0}\frac{\partial A_z(R_{w-})}{\partial \rho} = K \end{cases} \tag{3.9}$$

μ_0 permeability of vacuum, $\mu_0 = 4\pi \times 10^{-7}$ A/m, it can be solved as:

$$A_z(\rho, \phi) = \begin{cases} \frac{\mu_0 K_m}{2p_0}\rho\left(\frac{\rho}{R_w}\right)^{p-1}\sin(p\phi) & \rho < R_w \\ \frac{\mu_0 K_m}{2p_0}\rho\left(\frac{R_w}{\rho}\right)^{p+1}\sin(p\phi) & \rho \geq R_w \end{cases} \tag{3.10}$$

According to the definition of magnetic vector potential, it is possible to obtain radial magnetic dense B_ρ and radial magnetic density B_ϕ at any point in the air-core pulsed alternator by,

$$B_\rho(\rho, \phi) = \frac{1}{\rho}\frac{\partial A_z(\rho, \phi)}{\partial \phi} = \begin{cases} \frac{\mu_0 K_m}{2}\left(\frac{\rho}{R_w}\right)^{p-1}\cos(p\phi) & \rho < R_w \\ \frac{\mu_0 K_m}{2}\left(\frac{R_w}{\rho}\right)^{p+1}\cos(p\phi) & \rho \geq R_w \end{cases} \tag{3.11}$$

$$B_\phi(\rho, \phi) = -\frac{\partial A_z(\rho, \phi)}{\partial \rho} = \begin{cases} -\frac{\mu_0 K_m}{2}\left(\frac{\rho}{R_w}\right)^{p-1}\sin(p\phi) & \rho < R_w \\ \frac{\mu_0 K_m}{2}\left(\frac{R_w}{\rho}\right)^{p+1}\sin(p\phi) & \rho \geq R_w \end{cases} \tag{3.12}$$

Because only excitation windings have current when there is no load, it can be known from the upper formula that at this time the excitation winding is of the maximum magnetic density. The external and internal magnetic density decreases with the increase of the distance to the winding. The internal one ($\rho < R_w$) decreases more slowly than the external one ($\rho > R_w$). The speed of attenuation depends on p_O, the number of the pole pairs. The larger the value of p_O is, the faster the attenuation will be. But from the aspect of electromagnetic shielding, the larger the value of p_O is, the more favorable it will be for the outer shield and the shaft shield of the motor. When the motor is bipolar, the magnetic density on each point of the excitation winding is equal, and the magnetic force line goes through the shaft, which requires the motor to adopt a non-magnetic shaft. That's why it's not common for the air-core pulsed alternators to be bipolar.

It can also be seen from the (3.11) that there is a linearly proportional relationship between the magnetic field and the excitation current in the air-core pulsed alternator because there is no magnetic saturation constraint. The radial and tangential components of the base wave air gap are of the same amplitude, which is completely different from the iron-core motor in which only radial components can be seen. It requires additional accounting of the effect of radial magnetic density to the radial electromagnetic force of the armature winding.

3.3.2 *Iron-Core Alternator*

When the motor is of iron-core topology, the effect on magnetic density is only multiplied by the corresponding coefficient on the basis of the calculation of the air-core magnetic density, as shown in Table 3.1.

Among them, λ_R and λ_S represent the magnetic permeability of the stator and the rotor respectively, satisfying (3.13) and (3.14). Its maximum value is 1 (at this time the differential permeability is infinite), and the minimum value is −1.0 (at this time the differential permeability is 0), as shown in Fig. 3.4.

$$\lambda_R = (\mu_r - 1)/(\mu_r + 1) \tag{3.13}$$

$$\lambda_S = (\mu_s - 1)/(\mu_s + 1) \tag{3.14}$$

As can be seen in Table 3.1, when the only stator or rotor is of iron-core, the air gap magnetic density of the rotor will be increased to twice as the circumstance in which both of them are air-core (regardless of saturation). It can be analyzed from the above that in terms of the magnetic circuit, when the stator or rotor is iron-core, the differential permeability is infinite, the magnetic resistance is approximately zero, which is equivalent to the magnetic short circuit of the stator or the rotor. The magnetic resistance of the entire motor is halved, so in the same magnetic potential, the magnetic density of the air gap will be increased to twice as the circumstance in which both of them are air-core. Correspondingly, when both the stator and the

Table 3.1 The laws of magnetic fields of generators pulsed by different materials

	$\rho < R_w$	$\rho > R_w$
Air-core rotor Air-core stator	$B_r = \frac{\mu_0 A}{2}\left(\frac{\rho}{R_w}\right)^{p-1} \cos p\phi$	$B_r = \frac{\mu_0 A}{2}\left(\frac{R_w}{\rho}\right)^{p+1} \cos p\phi$
Iron-core rotor Air-core stator	$B_r = \frac{\mu_0 A}{2}\left(\frac{\rho}{R_w}\right)^{p-1}\left[1+\lambda_R\left(\frac{R_r}{\rho}\right)^{2p}\right] \cos p\phi$	$B_r = \frac{\mu_0 A}{2}\left(\frac{R_w}{\rho}\right)^{p+1}\left[1+\lambda_R\left(\frac{R_r}{R_w}\right)^{2p}\right] \cos p\phi$
Iron-core stator Air-core rotor	$B_r = \frac{\mu_0 A}{2}\left(\frac{\rho}{R_w}\right)^{p-1}\left[1+\lambda_s\left(\frac{R_r}{R_s}\right)^{2p}\right] \cos p\phi$	$B_r = \frac{\mu_0 A}{2}\left(\frac{R_w}{r}\right)^{p+1}\left[1+\lambda_s\left(\frac{\rho}{R_s}\right)^{2p}\right] \cos p\phi$
Iron-core rotor Iron-core stator	$\rho < R_w$时, $B_r = \frac{\mu_0 A}{2}\left(\frac{\rho}{R_w}\right)^{p-1} \frac{\left[1+\lambda_R\left(\frac{R_r}{R_w}\right)^{2p}\right]\left[1+\lambda_S\left(\frac{\rho}{R_s}\right)^{2p}\right]}{\left[1-\lambda_R\lambda_S\left(\frac{R_r}{R_s}\right)^{2p}\right]} \cos p\phi$ $\rho > R_w$时, $B_r = \frac{\mu_0 A}{2}\left(\frac{R_w}{\rho}\right)^{p+1} \frac{\left[1+\lambda_R\left(\frac{R_r}{R_w}\right)^{2p}\right]\left[1+\lambda_S\left(\frac{\rho}{R_s}\right)^{2p}\right]}{\left[1-\lambda_R\lambda_S\left(\frac{R_r}{R_s}\right)^{2p}\right]} \cos p\phi$	

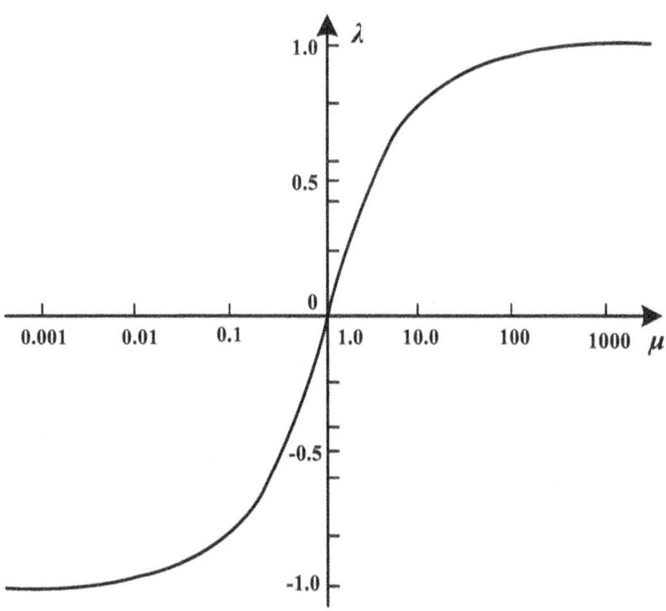

Fig. 3.4 Relationship between the magnetic permeability and differential permeability

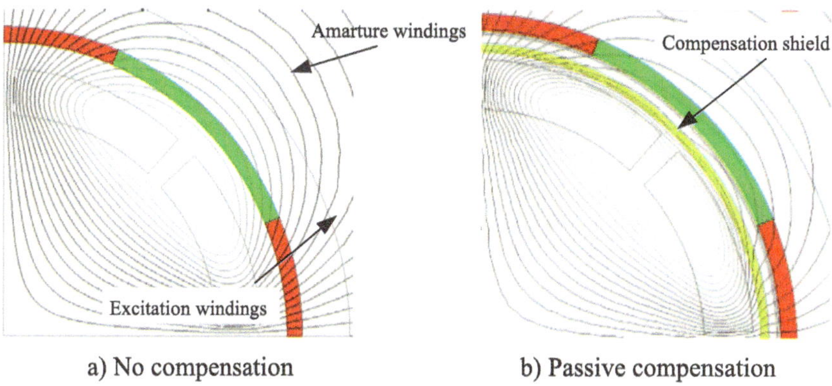

Fig. 3.5 Distribution of magnetic force lines during self-stimulation

rotor are iron-core, the magnetic density of the motor air gap can be increased to four times as the circumstance in which both of them are air-core.

Since the pulsed alternator is of an air-core structure, the magnetic field is dispersed around the motor, the electromagnetic shielding needs to be considered. Ideally, it would be assumed that the shielded part has no magnetic field at all, which means that the differential permeability is zero. As can be seen from Fig. 3.4, $\lambda = -1$, when the case is shielded, it can be seen in the formula of the rotor of air-core and the stator of iron-core that the shield weakens the magnetic field on the outside of the excitation winding. This suggests that the closer it is to the shield, the closer the radial magnetic dense is to zero. Therefore, when designing shielding, the researchers need to try to stay away from excitation windings, and also consider not to over-increase the volume of the motor.

When the pulsed alternator adopts a passive compensation structure, the shielded aluminum cylinder will be added in the middle of the stator and the rotor. As can be seen from Table 3.1, the shielded part has no magnetic field, which means that the differential permeability is zero. As can be seen from Fig. 3.4, when the shield is ideal, $\lambda = -1$. This implies that the closer the alternator is to the shield, the closer to zero the radial magnetic density will be. The magnetic density only has tangential components, as shown in Fig. 3.5.

3.4 Calculation of Key Parameters of Pulsed Alternators

For the pulsed alternator, the two most important parameters are inductance and coupling coefficient, which are directly related to the amount of the output pulsed power, which is of great significance to the motor design. According to the law of the motor field distribution discussed in the previous section, the inductance of the winding can be calculated by energy method, and the magnetic energy per unit length is

$$W_m = \frac{1}{2}\int_l (\mathbf{A}\cdot\mathbf{K})dl \tag{3.15}$$

The expression of the general self-inductance of the winding L is

$$\begin{aligned} L &= \frac{2W_m}{i^2} = 2p\cdot\frac{1}{i^2}\cdot\int_0^{\frac{\pi}{p}} A_z(\rho)\cdot K\cdot\rho d\phi \\ &= 2p\cdot\frac{1}{i^2}\cdot\int_0^{\frac{\pi}{p}} \frac{\mu_0 K_m}{2p}\rho\sin(p\phi)\cdot K_m\sin(p\phi)\cdot\rho d\phi \\ &= \frac{2\mu_0}{\pi}\frac{(Nk_w)^2}{p} \end{aligned} \tag{3.16}$$

For air-core pulsed alternators, L can be used to represent the self-inductance of the excitation winding (L_f), self-inductance of armature winding (L_a), and self-inductance of compensation winding (L_c). With the same method, the expression of general mutual inductance M can be derived, assuming that the average radius of the two sets of windings are r_1 *and* r_2 respectively, (the value of r_1 is less than r_2), and the mechanical angle of the winding axis is σ. The expression of M is

$$\begin{aligned} M_{12} &= 2p\cdot\frac{1}{i_1 i_2}\cdot\int_0^{\frac{\pi}{p}} A_z(r_1)\cdot K_2\cdot r_1 d\phi \\ &= 2p\cdot\frac{1}{i_1 i_2}\cdot\int_0^{\frac{\pi}{p}} \frac{\mu_0 K_{m1}}{2p_0} r_1\left(\frac{r_1}{r_2}\right)^{p_0-1} \sin(p_0\phi)\cdot K_{m2}\sin(p(\phi-\sigma))\cdot r_1 d\phi \\ &= \frac{2\mu_0}{\pi}\frac{(N_1k_{w1})(N_2k_{w2})}{p}\left(\frac{r_1}{r_2}\right)^{p_0}\cos(p\sigma) \end{aligned} \tag{3.17}$$

M_{12} can be used to represent the mutual inductance between the armature winding and excitation winding (M_{af}), the armature winding and compensation winding (M_{ac}), and the compensation winding and excitation winding (M_{cf}).

Coupling coefficient is one of the key parameters of the pulsed alternator, when the axial angle between the rotor and stator (σ) is 0, the value of the universal coupling coefficient k is the largest, it can be expressed as

$$k = \frac{M_{12}}{\sqrt{L_1 L_2}} = \left(\frac{r_1}{r_2}\right)^p \tag{3.18}$$

To achieve a lower equivalent inductance for armature winding and a faster self-excitation process, the pulsed alternator design requires the coupling coefficient of the armature winding and compensation windings and excitation windings to be as large as possible. Therefore, only from the theoretical point of view, the excitation windings and compensation windings should be as close as possible to the armature windings, and the number of the poles of the motor should be as small as possible. In practical applications, limitations such as structural strength and manufacturing processes also need to be considered.

When the motor is not of total air-core structure, according to the distribution law of magnetic density discussed in Table 3.1 above, at this time when calculating the inductance only the corresponding coefficient need to be multiplied.

In particular, when the case is subject to electromagnetic shielding, the expression of the armature winding inductance can be acquired according to the law from the above section as

$$L_{asd} = \frac{2\mu_0}{\pi}\frac{(N_a k_{wa})^2}{p}\left[1-\left(\frac{r_a}{r_{sd}}\right)^{2p}\right] \tag{3.19}$$

r_a the average radius of the armature winding;
r_{sd} radius of the casing.

From the formula, such gain can be figured out that shielding decreases the inductance of the armature windings.

The expressions of mutual inductance of N_a(the turns of armature windings) and the N_f(the turns of excitation windings) as:

$$M_{afsd} = \frac{2\mu_0}{\pi}\frac{(N_a k_{wa})(N_f k_{wf})}{p}\left(\frac{r_a}{r_f}\right)^p\left[1-\left(\frac{r_a}{r_{sd}}\right)^{2p}\right] \tag{3.20}$$

r_f the average radius at the excitation winding position;
k_{wa} the base wave coefficient of the armature winding;
k_{wf} the base wave coefficient of the excitation winding.

From the formula, the shielding reduces the mutual inductance between the armature windings and the excitation windings.

From the above analysis, it can be seen that while the electromagnetic shield provides an effective shield for the pulsed alternator, the reaction of the eddy current magnetic field of the shield also changes the magnetic field distribution in the motor. On the one hand, it reduces the self-inductance of the armature winding, which is conducive to improving the output power of the motor. However, on the other hand, the reduction of the mutual inductance of the armature winding and excitation winding also affects the electromotive force coefficient, which in return reduces the armature induction voltage generated by the unit excitation current, thus decreasing the output voltage of the motor, delaying the magnetic self-excitation process, and lowering the power and efficiency of the motor.

3.5 Analysis of Discharge Characteristics of Pulsed Alternators

3.5.1 Analysis of the Process of the Discharging of Pulsed Alternators

The discharging process of the pulsed alternators to the loads (very small ones) can be compared to the sudden short-circuiting process of a conventional motor. According to the theory of a sudden short circuit in the three-phase synchronous generator, the short-circuit process follows the following formula

$$i \approx i_{\sim} + i_{=} = E_m\left[\frac{1}{X} + \left(\frac{1}{X'} - \frac{1}{X}\right)e^{-\frac{t}{T'}} + \left(\frac{1}{X''} - \frac{1}{X'}\right)e^{-\frac{t}{T''}}\right]\cos(\omega t + \theta_0) - \frac{E_m}{X''}\cos\theta_0 e^{-\frac{t}{T_a}} \tag{3.21}$$

Wherein, the short-circuit current has both the period component and the non-periodic DC attenuation component, and the existence of the non-periodic DC component depends on the initial angle of the rotor when the circuit is shortened When a pulsed alternator drives an electromagnetic load, it generally requires a smooth waveform, so the existence of DC attenuation current is not preferable. In addition, in order to obtain a current waveform with a short rise time, the trigger angle is set to be 90°, at which point the non-periodic component can be eliminated.

During the discharge of the pulsed alternator, it's necessary to pay special attention to the value of the current peak. Since the load is purely perceptual, the current change lags behind the voltage by 90°, so the maximum current will appear only in a quarter of the cycle. Due to the short time interval, the exponential attenuation item is approximately 1.0. The approximate calculation formula of the discharge current peak is:

$$i_m \approx \frac{E_m}{\omega\left(1 - k^2\right)L_a + \omega L_x} \tag{3.22}$$

The above formula can only approximate the process of discharge of the pulsed alternator. Strictly speaking, the discharge of pulsed alternators is an asymmetric and sudden short-circuit process, and there is a compensation structure, so it is necessary to polish the symmetric short-circuit theory of the synchronous motor. The pulsed alternators can be classified into two categories, namely compensation structure and non-compensation structure. This article will analyze them separately.

1. Compensation structure

When there is a compensation structure, no matter what kind of structure–active compensation, passive compensation, or passive alternation, there is generally a cross-axis compensation part, and excitation windings are equivalent to straight shaft compensation. Therefore, the existence of compensation structure comes with straight shaft compensation and cross-axis compensation.

When the pulsed alternator discharges, the compensation structure can reduce the armature winding inductance on the one hand, and shield the magnetic field generated by the discharge current on the other hand to avoid the excitation windings from being exposed to the strong magnetic field. Exposed to a strong discharge magnetic field, there will be a significant loss in the excitation windings, and the overvoltage will be induced by the armature windings, which will affect the motor insulation and switching devices. During the discharge process, the discharge voltage is basically unchanged, and the discharge current can be expressed in the Eq. (3.22). Due to the constant changes in load impedance and armature transient impedance, the maximum current value sits at different pulses.

2. No compensation structure

In this case, the motor has no compensation structure, and the excitation winding acts as a magnetic field and direct-axis compensation. Without compensation structure, the discharge magnetic field penetrates the excitation winding, which senses high voltage, and in turn affects the next discharge process. Therefore, the discharge process can not simply be described in the same way as the three-phase symmetrical short circuits.

At this time, there are two stages in the discharge process. This paper takes the two-phase orthogonal pulsed alternator as an example for analysis.

Stage 1:

In order to eliminate the non-periodic component in the discharge current, the pulsed alternator typically triggers a discharge at a voltage of 90°, as shown in Fig. 3.6a. After a quarter of a cycle, the rotor goes to the position shown in Fig. 3.6b. In phase A the voltage is 0, in phase B voltage reaches maximum. The load is purely perceptual,

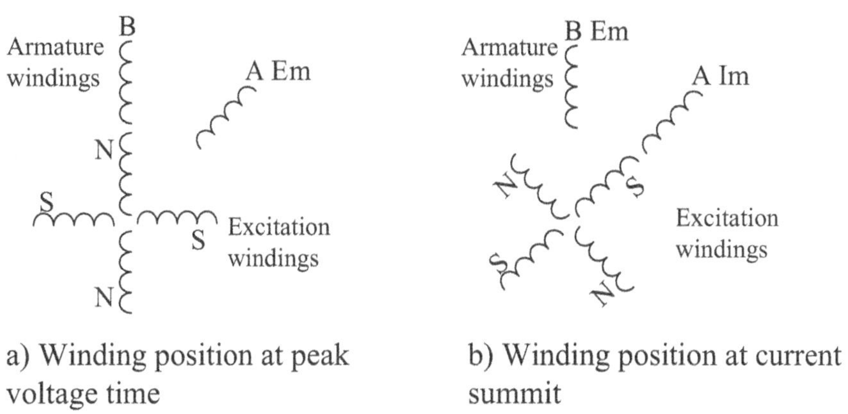

Fig. 3.6 Changes in winding positions during discharge

and the current lags behind the voltage by 90°. Phase A current reaches its peak value I_m and can be calculated according to the formula (3.22). At this point, the interaction M_{af} between the phase A winding and the excitation winding is maximized.

$$L_f \Delta i_f = M_{af} i \tag{3.23}$$

According to the conservation law of circuit magnetic flux, as shown in the formula (3.23), there is the maximum Δi_f. Because phase A and B are orthogonal, excitation winding and the magnetic flux on phase B hinge is not affected by the demagnetized flux on phase A, the magnetic density on phase B will increase, while the anti-electric potential will rise to $(\Delta i_f + i_f)E_m/i_f$.

Stage 2:

After another quarter of a cycle, the rotor of phase B turns over the 90° electrical angle, which makes the corresponding current reaches its peak value. Because in the previous stage, when phase B triggered the release point, the phase B voltage increased, in this turn the phase B current peak is also multiplied by the corresponding times to $(\Delta i_f + i_f)E_m/i_f$.

The above process repeats. In order to study on the trends in the armature voltage and discharge current during discharge, the equations for each pulse are listed below:

First pulse:

$$i_1 = \frac{E_m}{\omega\left(1 - k^2\right)L_a + \omega L_x} \tag{3.24}$$

$$L_f \Delta i_{f1} = M_m i_1 \tag{3.25}$$

Second pulse:

$$i_2 = \frac{\Delta i_{f1} + i_f}{i_f} i_1 = i_1 + \frac{M_{af}}{L_f i_f} i_1^2 \tag{3.26}$$

$$L_f \Delta i_{f2} = M_m i_2 \tag{3.27}$$

Third pulse:

$$i_2 = \frac{\Delta i_{f2} + i_f}{i_f} i_1 = i_1 + \frac{M_{af}}{L_f i_f} i_1^2 + \frac{M_{af}^2}{L_f^2 i_f^2} i_1^3 \tag{3.28}$$

$$L_f \Delta i_{f3} = M_m i_3 \tag{3.29}$$

And so on, the nth pulse:

$$i_n = \frac{\Delta i_{fn} + i_f}{i_f} i_1 = i_1 + \frac{M_m}{L_f i_f} i_1^2 + \frac{M_m^2}{L_f^2 i_f^2} i_1^3 + \cdots + \frac{M_m^{n-1}}{L_f^{n-1} i_f^{n-1}} i_1^n \tag{3.30}$$

Formula (3.30) meets the summation of geometric progression $q = M_m i_1 / L_f\ i_f$, therefore

$$i_n = \frac{i_1(1 - q^n)}{1 - q} \tag{3.31}$$

If q is less than 1, after several pulses, the peak value of the armature voltage and discharge current does not rise and the discharge process stabilizes.

When a no-compensated pulsed alternator discharges, the armature winding voltage and discharge current will increase gradually, affecting insulation and switching devices. However, it has a positive effect on high-voltage applications and pulsed current waveform regulation.

(1) High-voltage application

Compared to capacitors, the pulsed power supply is with low voltage and high current, which is suitable for driving orbital gun loads. Lasers and microwave generators require high voltage, therefore conventional compensated pulsed alternators are not suitable for such loads. According to the above principle, the compensation structure is not favorable for pulsed alternator, so that the discharge magnetic field can penetrate the excitation winding, and the high voltage can be induced on the armature winding therefore driving this kind of load.

(2) Electromagnetic emission waveform adjustment

When the pulsed alternator drives the load of the orbital gun, the discharge current is gradually decayed and the acceleration of projectile is gradually reduced, resulting in a relatively low acceleration of orbital gun emission due to the increasing load impedance and anti-electric potential. In order to make the discharge current satisfactory the traditional method is to use different forms of compensation structure or connect the rotation flux compressor in series in the pulsed alternator by adjusting the internal impedance of the power supply in order to balance the increasing load impedance, so that the total impedance in the circuit is approximately unchanged, and a flat discharge current will be obtained.

The no-compensated pulsed alternator studied in this section, the increasing discharge characteristic of the armature voltage, can resist the effect of load impedance and anti-electric potential. After detailed calculation and design, it can be drawn that when the difference between the armature voltage of the pulsed alternator and the counter-electric potential of the load is proportional to the load impedance, the discharge current can be guaranteed to be flat-top, and the projectile gains a better acceleration ratio.

3.5.2 Analysis of Factors Affecting the Discharge Current of Pulsed Alternators

The compensation of pulsed alternators is divided into three forms: active compensation, passive compensation and passive alternation compensation. This paper takes passive compensated pulsed alternator as an example because of the following reasons, i.e. its structure is simple, its discharge current waveform is similar to the sine, and it can be widely used. The analysis of the factors that affect the discharge current of the pulsed alternator in this paper is based on the circuit theory.

As shown in Fig. 3.7 is how the passive compensated pulsed alternator of separate excitation and transition type drives the load in a simplified circuit modelling.

When discharging, the spatial magnetic field in which the compensation cylinder located changes, and the potential induced in the compensation cylinder produces a eddy current, thus reducing the compensation cylinder into a circuit in series of resistance and inductance. The discharge circuit and the compensation circuit fit with the convention, while the excitation circuit suits the motor. The circuit equations for discharge loops, compensation loops, and excitation circuits are:

$$\begin{cases} E_a = L_a\frac{di_a}{dt} + M_{ac}\frac{di_c}{dt} + M_{af}\frac{di_f}{dt} - R_a i_a - R_l i_a \\ 0 = L_c\frac{di_c}{dt} + M_{ca}\frac{di_a}{dt} + M_{cf}\frac{di_f}{dt} - R_c i_c \\ V_{DC} = L_f\frac{di_f}{dt} + M_{fa}\frac{di_a}{dt} + M_{fc}\frac{di_c}{dt} + R_f i_f \end{cases} \tag{3.32}$$

E_a the output voltage of the pulsed alternator;
V_{DC} DC excitation voltage;
i_f excitation current;
i_a discharge current of the armature winding;
i_c the inductant eddy current in the cylinder;

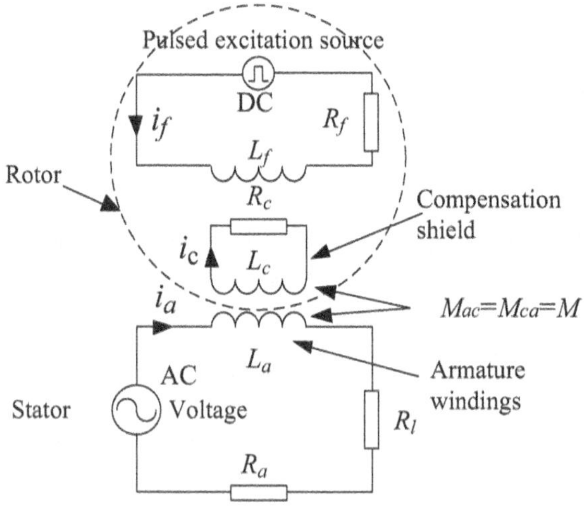

Fig. 3.7 Simplified circuit model for passive compensated pulsed alternators

R_l the resistance of the pulsed load;
L_a, R_a the self-inductance and resistance of armature windings;
L_f, R_f the self-inductance and resistance of excitation windings;
L_c, R_c the equivalent self-inductance and resistance of compensation cylinder;
M_{ac}, M_{ca} the mutual inductance of discharge circuit and compensation circuit;
M_{af}, M_{fa} the mutual inductance of discharge circuit and excitation circuit;
M_{cf}, M_{fc} the mutual inductance of excitation circuit and compensation circuit.

If the DC pulsed supply excitation is used, di_f/dt is equal to zero, the influence of the excitation circuit on the discharge circuit and the compensation circuit can be ignored. The effect of the compensation element on the discharge current is analyzed here, so the effect of the armature winding sand and the compensation loop on the excitation winding during discharge is not considered. In a compensated pulsed alternator, the internal resistance of the armature winding and the equivalent resistance of the compensation circuit are usually below the milliohm level and can be ignored. While ignoring the resistance of the pulsed load, the circuit equation for the discharge circuit and the compensation circuit the Eq. (3.32) can be reduced to:

$$\begin{cases} E_a = L_a \frac{di_a}{dt} + M_{ac} \frac{di_c}{dt} \\ 0 = L_a \frac{di_a}{dt} + M_{ac} \frac{di_c}{dt} \end{cases} \tag{3.33}$$

In the equation, the mutual inductance of the discharge circuit M_{ac} *and* M_{ca} are equal and can be recorded as M. The rate of change of the pulsed current output over time is:

$$\frac{di_a}{dt} = \frac{E_a}{L_a - \frac{M^2}{L_c}} \tag{3.34}$$

According to the definition of the coupling coefficient, the coupling coefficient k for the discharge and compensation circuits is:

$$k = \frac{M}{\sqrt{L_a L_c}} \tag{3.35}$$

The formula (3.34) can be further rewritten as:

$$\frac{di_a}{dt} = \frac{E_a}{L_a(1 - k^2)} \tag{3.36}$$

If the change in speed is taken into account, the instantaneous value of the anti-electric potential output of the pulsed alternator is:

$$E_a = \frac{E_m n}{n_0} \sin(\omega t) \tag{3.37}$$

E_m the peak value of the anti-electric potential in armature windings of the motor $E_m = Np\Phi n_0/30\alpha_p$;
N the number of turns per pole conductor;
P the number of the pair of poles;
Φ the total magnetic flux per pole;
n_0 rated speed;
α_p pole arc coefficient.

Due to the non-conductive magnetic material inside air-core pulsed alternator, the total magnetic flux per pole is proportional to the excitation current.

$$\Phi = K_f i_f \tag{3.38}$$

Bring formula (3.38) into Eq. (3.37) and convert the rotor speed n to the rotation angle speed ω, hereby:

$$E_m = \beta\omega i_f \tag{3.39}$$

β generator consonant;
ω rotational angular velocity.

In order to meet the load requirements, it is necessary to make the change rate di_a/dt of pulsed current output of the pulsed alternators as big as possible. From formula (3.36) it can be seen that in order to improve the discharge current change rate di_a/dt, it is necessary to increase the generator's anti-electric potential peak value E_m under no load, and reduce the inductance of armature winding L_a, while increasing the coupling coefficient k of the armature winding and compensation winding (compensation cylinder).

When the size of the motor remains unchanged, to improve the alternators' anti-electric potential E_a under no load mainly relies on the increase of the rotation speed n, the air gap magnetic density B and the number of the turns of windings N. The increase in the number of turns will greatly increase the inductance of the winding, so it's favorable to adopt the first two methods. High speeds require high-strength materials to overcome the centrifugal forces generated by high speeds and high-speed bearings. For the air-core compensated pulsed alternator, to get a higher magnetic density, it mainly depends on self-excitation.

3.5.3 Analysis of the Conditions for Self-Excitation Establishment of Air-Core Pulsed Alternators

For motors made of conventional ferromagnetic material, self-excitation magnetism is usually not adopted because of excessive excitation current. It is highly possible to burn the motor winding. At the same time, it tends to be supersaturated and difficult

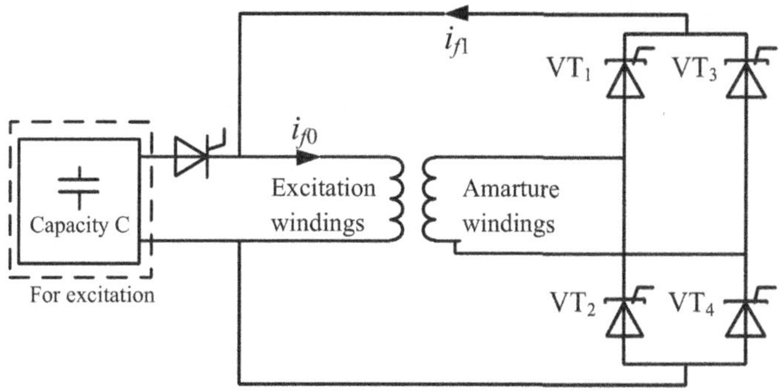

Fig. 3.8 The circuit wiring diagram of compensated pulsed alternators

to control. The yokes of the rotors and stators inside the air-core pulsed alternators are made of non-magnetic fiber epoxy composite material, and the magnetic potential required by the magnetic field for the motor is several times higher than the iron-core motor. As there is no saturation phenomenon, the self-excitation process is short, so self-stimulation can be used. The circuit diagram is shown in Fig. 3.8. In operation, a small seed current is injected into the excitation winding, and after the output voltage of the armature winding is repelled, it is received to the excitation rotor by the electric brush slip ring, and the required excitation current is obtained through the positive feedback process. Positive feedback process are not be established in all the cases, except when certain conditions are met. The realization of self-excitation is directly related to the design of the motor, so it is necessary to study on the approaches to self-excitation in order to guide the design of pulsed alternators.

The equation for the excitation feedback circuit is:

$$L_f \frac{di_f}{dt} + R_f i_f = E_C \tag{3.40}$$

E_c DC voltage after armature anti-electric potential rectification, ignoring the armature winding resistance R_a, $E_c = c\omega i_f$;

C Ratio of the output voltage after rectification to the effective value of the armature anti-electric potential.

Therefore the self-excitation current i_f should be:

$$i_f = I_0 e^{\left(\frac{c\omega - R_f}{L_f}\right)t} \tag{3.41}$$

I_0 the seed current provided to the excitation windings from the excitation capacitor.

The prerequisite of a successful self-excitation is that the current from the self-excitation feedback i_{f1} should be greater than the excitation current i_{f0},

$$i_{f1} > i_{f0} = I_0 \tag{3.42}$$

i_{f1}, i_{f0} as shown in Fig. 3.8. The prerequisites are:

$$c\omega > R_f \tag{3.43}$$

At the same time, if armature reactions are considered, the conditions for a successful self-excitation can be derived from the simplified model:

$$\omega M > 1.57\left(R_a + R_f\right) \tag{3.44}$$

M the maximum value for the mutual inductance of the armature and excitation windings.

In the design of the motor, the formula (3.44) must be used to verify whether the self-excitation condition is valid. Therefore, the target of the electromagnetic design for the core pulsed alternator can be attributed to high speed, high mutual inductance and low internal resistance. High speed sits not only to improve the energy storage and power density of the motor, but also to help achieve self-excitation. In design, the strength of the structure and insulation need to be ensured while reducing the radial distance between the armature winding and excitation winding and increasing the electromagnetic coupling between the armature winding and excitation winding. This is beneficial not only to the achievement of self-excitation, but also to higher efficiency of the motor. The low internal resistance can lower the winding copper consumption of the motor. In this way the motor efficiency is improved, and the demand of high peak output current is fulfilled. The self-excitation circuit of the air-core pulsed alternator adopts the full-bridge rectifier structure to obtain a larger c value and accelerate self-excitation.

Self-excitation is also used in traditional iron-core pulsed alternators. Due to the saturation characteristics of the iron-core motor, with the increase of self-excitation current, the saturation of the motor increases. In addition, the voltage increasement of the armature winding also tends to saturate, and eventually the voltage will be equal to the voltage at both ends of the excitation winding and will automatically reach the end point. The self-excitations of air-core and iron-core rotor are different. Because the internal material is linear, as long as the excitation current increases, the armature winding voltage will increase proportionally, thus the endpoint must be artificially controlled. After reaching the required armature winding voltage, through the rectifier phase control, the excitation current will be constant for a period of time. After discharging completed, the remaining energy in the armature winding is then fed back to the excitation winding by the control rectifier, thus accelerates the motor rotor. The air-core pulsed alternator's self-discharge process is shown in Fig. 3.9.

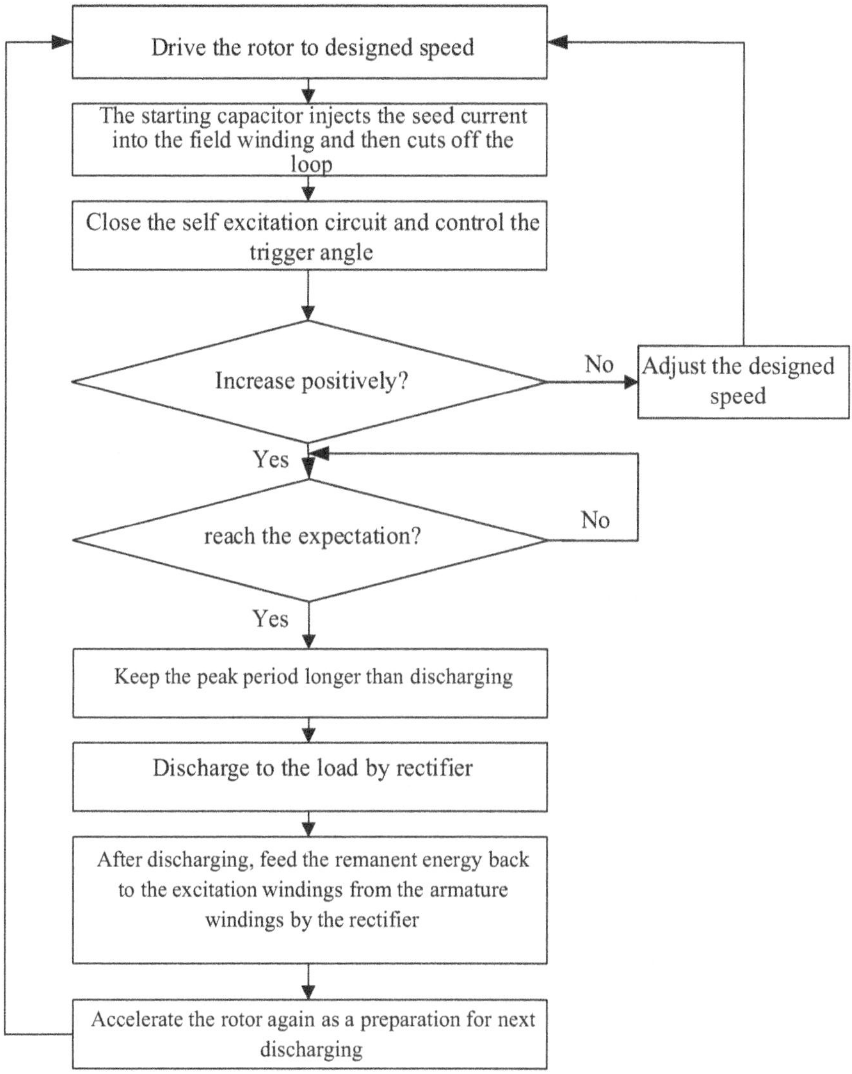

Fig. 3.9 The discharging flowchart of air-core compensated pulsed alternator

3.6 Mathematical Model of Pulsed Alternators

The discharging process of pulsed alternator is extremely complex. In addition to considering the effect of the generator's instantaneous short circuit on the motor, it is also necessary to consider the effect of the change of speed during the inertial rotation of the motor on the discharge process. Section 3.5 studies on the discharge process of pulsed alternator. The study only described the peak value of current while ignored the effect of rotation speed and load changes. In order to describe the discharge current more accurately, a mathematical model of the pulsed alternator is to be established.

3.6.1 Mathematical Model of the Air-Core Pulsed Alternators in the Phase Coordinate System

Referring to the mathematical model of synchronous motor, a four-phase air-core compensated pulsed alternator with a cross-axis compensation winding is selected as an example, and a mathematical model of the air-core pulsed alternator is established as shown in Fig. 3.10.

The air-core pulsed alternator is of a rotating magnetic field structure, while the mutual difference among stator four-phase armature winding is an electrical angle of 90°. The rotor is equipped with a short circuit cross-axis compensation winding, and orthogonal with the excitation winding. Because the fiber epoxy composite material is not featured with magnetic saturation, the magnetic circuit of the motor is linear. This meets the assumptions of the ideal motor. In order to study the basic electromagnetic relationship inside the air-core pulsed alternator, it is necessary to make the following assumptions:

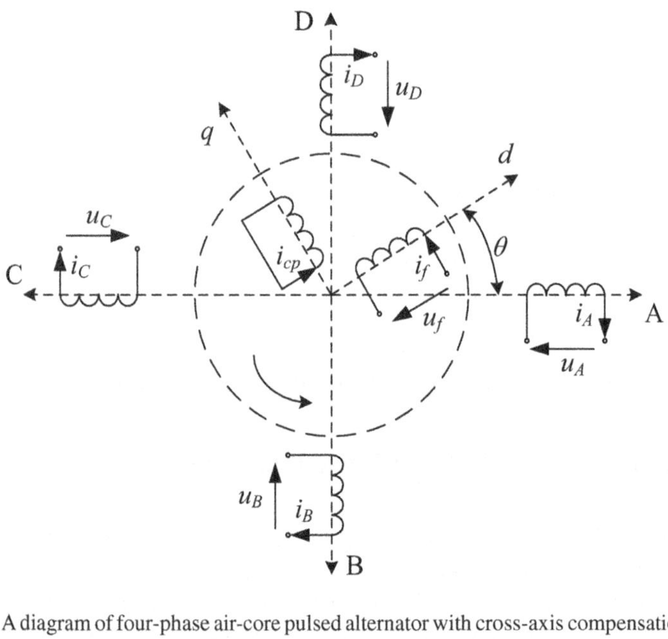

Fig. 3.10 A diagram of four-phase air-core pulsed alternator with cross-axis compensation winding

1. Ignoring the higher harmonics of the magnetic field, it is believed that the air gap magnetic field is in sine distribution in space. When a harmonic magnetic field needs to be considered, the concept of harmonic leakage resistance can be introduced to calculate its effects.
2. Avoiding the consideration of the effect of the extra-stator electromagnetic shield on the electromagnetic relationship within the motor. When considering the shielding effect, it is required to produce two equivalent sets of stationary orthogonal short-circuit windings outside the stator and add two state variables to the state equation.

The positive directions of voltage, current, magnetic chain and torque in the mathematical model are shown as follows:

1. The stator armature winding adopts the convention of generators by taking the direction of the output current as positive. Each phase of winding flows through the positive current to produce a negative magnetic chain.
2. The rotor excitation windings and the cross-axis compensation windings adopts the convention of motors by taking the direction of the output current as positive. Each phase of winding flows through the positive current to produce a positive magnetic chain.

Under the above-mentioned positive directions, the magnetic chain equation of a four-phase air-core pulsed alternator can be written in the form of a matrix:

$$\begin{pmatrix} \boldsymbol{\Psi}_s \\ \boldsymbol{\Psi}_r \end{pmatrix} = \begin{pmatrix} \boldsymbol{L}_s & \mathbf{M}_{sr} \\ \boldsymbol{M}_{rs} & \boldsymbol{L}_r \end{pmatrix} \begin{pmatrix} -\boldsymbol{i}_s \\ \boldsymbol{i}_r \end{pmatrix} \tag{3.45}$$

$\boldsymbol{\Psi}_s$ and $\boldsymbol{\Psi}_r$ represent the magnetic chain matrix of the stator and the rotor respectively:

$$\boldsymbol{\Psi}_s = \begin{pmatrix} \psi_A \ \psi_B \ \psi_C \ \psi_D \end{pmatrix}^T, \quad \boldsymbol{\Psi}_r = \begin{pmatrix} \psi_f \ \psi_{cp} \end{pmatrix}^T \tag{3.46}$$

$\boldsymbol{i}_s$ and $\boldsymbol{i}_r$ represent the current matrix of the stator and the rotor:

$$\boldsymbol{i}_s = \begin{pmatrix} i_A \ i_B \ i_C \ i_D \end{pmatrix}^T, \quad \boldsymbol{i}_r = \begin{pmatrix} i_f \ i_{cp} \end{pmatrix}^T \tag{3.47}$$

$\boldsymbol{L}_s$ and $\boldsymbol{L}_r$ represent the self inducted matrix of the stator and rotor windings respectively:

$$\boldsymbol{L}_s = \begin{pmatrix} L_{AA} & M_{AB} & M_{AC} & M_{AD} \\ M_{BA} & L_{BB} & M_{BC} & M_{BD} \\ M_{CA} & M_{CB} & L_{CC} & M_{CD} \\ M_{DA} & M_{DB} & M_{DC} & L_{DD} \end{pmatrix} = \begin{pmatrix} L_p & 0 & -M_p & 0 \\ 0 & L_p & 0 & -M_p \\ -M_p & 0 & L_p & 0 \\ 0 & -M_p & 0 & L_p \end{pmatrix} \tag{3.48}$$

$$\boldsymbol{L}_r = \begin{pmatrix} L_f & M_{fcp} \\ M_{cpf} & L_{cp} \end{pmatrix} = \begin{pmatrix} L_f & 0 \\ 0 & L_{cp} \end{pmatrix} \tag{3.49}$$

L_p the self-inductance of each phase winding of the stator;
M_p the mutual inductance between phase-A and phase-C, and phase-B and phase-D of the stator
L_f the self-inductance of the excitation windings of the rotor;
L_{cp} the self-inductance of the compensation windings of the rotor.

$\boldsymbol{M}_{sr}$ and $\boldsymbol{M}_{rs}$ show the mutual inductance matrix between the stator winding and the rotor winding respectively.

$$\boldsymbol{M}_{sr} = \boldsymbol{M}_{rs}^T = \begin{pmatrix} M_{Af} & M_{Acp} \\ M_{Bf} & M_{Bcp} \\ M_{Cf} & M_{Ccp} \\ M_{Df} & M_{Dcp} \end{pmatrix} = \begin{pmatrix} M_{fm}\cos\theta & -M_{cpm}\sin\theta \\ -M_{fm}\sin\theta & -M_{cpm}\cos\theta \\ -M_{fm}\cos\theta & M_{cpm}\sin\theta \\ M_{fm}\sin\theta & M_{cpm}\cos\theta \end{pmatrix} \tag{3.50}$$

M_{fm} the mutual inductance between the phase winding of the stator and the excitation winding of the rotor;
M_{cpm} the mutual inductance between the phase winding of the stator and the compensation winding;
θ The electrical angle between the rotor excitation winding and the axis of phase A of the stator.

According to Kirchhoff's second law, based on the prescribed positive direction, the voltage equation of the four-phase air-core pulsed alternator can be listed in the form of a matrix:

$$\begin{pmatrix} \boldsymbol{u}_s \\ \boldsymbol{u}_r \end{pmatrix} = \mathrm{p}\begin{pmatrix} \boldsymbol{\Psi}_s \\ \boldsymbol{\Psi}_r \end{pmatrix} + \begin{pmatrix} \boldsymbol{R}_s & \boldsymbol{O} \\ \boldsymbol{O} & \boldsymbol{R}_r \end{pmatrix}\begin{pmatrix} -\boldsymbol{i}_s \\ \boldsymbol{i}_r \end{pmatrix} \tag{3.51}$$

p differential operators in time domain $\mathrm{p} = d/dt$;

$\boldsymbol{u}_s$ and $\boldsymbol{u}_r$ represent the voltage matrix of the stator and the rotor one-to-one, i.e.

$$\boldsymbol{u}_s = \begin{pmatrix} u_A & u_B & u_C & u_D \end{pmatrix}^T, \boldsymbol{u}_r = \begin{pmatrix} u_f & u_{cp} \end{pmatrix}^T = \begin{pmatrix} u_f & 0 \end{pmatrix}^T \tag{3.52}$$

$\boldsymbol{R}_s$ and $\boldsymbol{R}_r$ represent the resistance matrix of the stator winding and the rotor winding indivilually, i.e.

$$\boldsymbol{R}_s = diag\begin{pmatrix} R_p & R_p & R_p & R_p \end{pmatrix}, \ \boldsymbol{R}_r = diag\begin{pmatrix} R_f & R_{cp} \end{pmatrix} \tag{3.53}$$

According to the virtual displacement method, the electromagnetic torque T_e of the four-phase air-core pulsed alternator is expressed as a matrix:

$$T_e = -\frac{p_0}{2}\begin{pmatrix} -\boldsymbol{i}_s & \boldsymbol{i}_r \end{pmatrix}\frac{\partial\begin{pmatrix} \boldsymbol{L}_s & \mathbf{M}_{sr} \\ \boldsymbol{M}_{rs} & \boldsymbol{L}_r \end{pmatrix}}{\partial\theta}\begin{pmatrix} -\boldsymbol{i}_s \\ \boldsymbol{i}_r \end{pmatrix} \tag{3.54}$$

By substituting the expressions of inductance to the above formula, it can be simplified to

$$\begin{aligned} T_e = {} & p_0 M_{fm} i_f [(i_A - i_C)\sin\theta + (i_D - i_B)\cos\theta] \\ & - p_0 M_{cpm} i_{cp} [(i_A - i_C)\cos\theta + (i_D - i_B)\sin\theta] \end{aligned} \tag{3.55}$$

According to Newton's second law, the torque equation for the pulsed alternator is:

$$T_1 = T_e + T_0 + J\frac{d\Omega}{dt} \tag{3.56}$$

T_1 the driving torque of the prime mover which is separated from the pulsed alternator before discharge. The pulsed alternator continues to rotate by inertia, at which time the driving torque is 0;

T_0 no-load resistance torque;

J the pulsed alternator's moment of inertia;

Ω the mechanical angular rate of the rotor.

3.6.2 Mathematical Model of Air-Core Pulsed Alternators Under the Rectangular Axis Coordinate System

As can be seen from the above section, under the phase coordinate system, the mutual inductance between the stator and rotor windings on the four-phase air-core pulsed alternators are both the functions of the angle θ, so the magnetic chain equation and voltage equation are differential equations containing time-varying coefficients, therefore it is difficult to attain the solution. Even if the circuit simulation model of analog square diagram on the computer doesn't help. In order to make the magnetic chain equation and voltage equation become linear constant coefficient differential equations, a $dqg0$ transformation need to be applied on the four-phase air-core pulsed alternator by reference to the Parker transformation of the three-phase synchronous generator.

The rotor excitation windings are orthogonal with the compensation windings and rotate synchronously with the rotor, so the rotor current does not need to be transformed, and the $dqg0$ transformation of the four-phase air-core pulsed alternator is only for the stator armature windings. As shown in Fig. 3.10, if projecting the four-phase current of the stator to the d-axis and q-axis that rotate with the rotor respectively, the stator current transformation matrix can be expressed as

$$\boldsymbol{i}_s = \boldsymbol{C}_{dqg0}\boldsymbol{i}'_s \; or \; \boldsymbol{i}'_s = \boldsymbol{C}_{dqg0}^{-1}\boldsymbol{i}_s \tag{3.57}$$

$\boldsymbol{i}'_s$ the stator current matrix after $dqg0$ transformation $\boldsymbol{i}'_s = \left(i_d \; i_q \; i_g \; i_0\right)^T$;

$\boldsymbol{C}_{dqg0}$ Coordinate transformation matrix, which satisfies the following conditions:

$$
C_{dqg0} = \begin{pmatrix} \frac{\cos\theta}{2} & -\frac{\sin\theta}{2} & 1 & 1 \\ -\frac{\sin\theta}{2} & -\frac{\cos\theta}{2} & -1 & 1 \\ -\frac{\cos\theta}{2} & \frac{\sin\theta}{2} & 1 & 1 \\ \frac{\sin\theta}{2} & \frac{\cos\theta}{2} & -1 & 1 \end{pmatrix} C_{dqg0}^{-1} = \begin{pmatrix} \cos\theta & -\sin\theta & -\cos\theta & \sin\theta \\ -\sin\theta & -\cos\theta & \sin\theta & \cos\theta \\ \frac{1}{4} & -\frac{1}{4} & \frac{1}{4} & -\frac{1}{4} \\ \frac{1}{4} & \frac{1}{4} & \frac{1}{4} & \frac{1}{4} \end{pmatrix} \tag{3.58}
$$

In the transformation, the g-axis and the zero axis are called semi-zero and zero-axis respectively. After the *dqg0* transformation, the magnetic chain equation of the four-phase air-core pulse alternator is:

$$
\begin{aligned}
\begin{pmatrix} \Psi_s' \\ \Psi_r' \end{pmatrix} &= \begin{pmatrix} C_{dqg0}^{-1} & O \\ O & I \end{pmatrix} \begin{pmatrix} L_s & \mathbf{M}_{sr} \\ M_{rs} & L_r \end{pmatrix} \begin{pmatrix} C_{dqg0} & O \\ O & I \end{pmatrix} \begin{pmatrix} -i'_s \\ i_r \end{pmatrix} \\
&= \begin{pmatrix} L'_s & \mathbf{M}'_{sr} \\ M'_{rs} & L'_r \end{pmatrix} \begin{pmatrix} -i'_s \\ i_r \end{pmatrix}
\end{aligned} \tag{3.59}
$$

By substitution to (3.48), (3.50) and (3.58), (3.59) is expanded to:

$$
\begin{pmatrix} \psi_A \\ \psi_B \\ \psi_C \\ \psi_D \\ \psi_f \\ \psi_{cp} \end{pmatrix} = \left(\begin{array}{cccc|cc} L_s & 0 & 0 & 0 & 2M_{fm} & 0 \\ 0 & L_s & 0 & 0 & 0 & 2M_{cpm} \\ 0 & 0 & L_o & 0 & 0 & 0 \\ 0 & 0 & 0 & L_o & 0 & 0 \\ \hline M_{fm} & 0 & 0 & 0 & L_f & 0 \\ 0 & M_{cpm} & 0 & 0 & 0 & L_{cp} \end{array}\right) \left(\begin{array}{c} -i_d \\ -i_q \\ -i_g \\ -i_0 \\ \hline i_f \\ i_{cp} \end{array}\right) \tag{3.60}
$$

The *dqg*0 transformation of the stator voltage equation of the formula (3.51) is performed. The voltage equation of the rotor remains the same, and the voltage equation of the four-phase air-core pulsed alternator after the transformation can be expressed as:

$$
\begin{aligned}
\begin{pmatrix} u'_s \\ u_r \end{pmatrix} &= \mathrm{p}\begin{pmatrix} \Psi_s' \\ \Psi_r \end{pmatrix} + \begin{pmatrix} \gamma'_s\omega_r\Psi_s' \\ O \end{pmatrix} + \begin{pmatrix} R'_s & O \\ O & R_r \end{pmatrix} \begin{pmatrix} -i'_s \\ i_r \end{pmatrix} \\
&= \left[\begin{pmatrix} L'_s & \mathbf{M}'_{sr} \\ M'_{rs} & L'_r \end{pmatrix} \mathrm{p} + \begin{pmatrix} \gamma'_s & O \\ O & O \end{pmatrix} \begin{pmatrix} L'_s & \mathbf{M}'_{sr} \\ M'_{rs} & L'_r \end{pmatrix} \omega_r + \begin{pmatrix} R'_s & O \\ O & R_r \end{pmatrix} \right] \begin{pmatrix} -i'_s \\ i_r \end{pmatrix}
\end{aligned} \tag{3.61}
$$

$$
\gamma'_s = C_{dqg0}^{-1} \frac{\partial C_{dqg0}}{d\theta} = \left(\begin{array}{cc|cc} 0 & -1 & 0 & 0 \\ 1 & 0 & 0 & 0 \\ \hline 0 & 0 & 0 & 0 \\ 0 & 0 & 0 & 0 \end{array}\right)
$$

ω_r the rotor's angular velocity (represented by electrical angle) $\omega_r = p_0\Omega$;
$\boldsymbol{R}_s$' Transformed resistance matrix, $\boldsymbol{R'}_s = \boldsymbol{C}_{dqg0}^{-1}\boldsymbol{R}_s\boldsymbol{C}_{dqg0} = \boldsymbol{R}_s$;

The formula (3.61) is expanded into the available voltage equation.

$$\begin{pmatrix} u_d \\ u_q \\ u_g \\ u_0 \\ \hline u_f \\ 0 \end{pmatrix} = \left(\begin{array}{cccc|cc} R_p + L_s\mathrm{p} & -L_s\omega_r & 0 & 0 & 2M_{fm}\mathrm{p} & -2M_{cpm}\omega_r \\ L_s\omega_r & R_p + L_s\mathrm{p} & 0 & 0 & 2M_{fm}\omega_r & 2M_{cpm}\mathrm{p} \\ 0 & 0 & R_p + L_o\mathrm{p} & 0 & 0 & 0 \\ 0 & 0 & 0 & R_p + L_o\mathrm{p} & 0 & 0 \\ \hline M_{fm}\mathrm{p} & 0 & 0 & 0 & R_f + L_f\mathrm{p} & 0 \\ 0 & M_{cpm}\mathrm{p} & 0 & 0 & 0 & R_{cp} + L_{cp}\mathrm{p} \end{array}\right) \begin{pmatrix} -i_d \\ -i_q \\ -i_g \\ -i_0 \\ \hline i_f \\ i_{cp} \end{pmatrix} \tag{3.62}$$

With substitution of (3.55) by i_d and i_q, the transformed electromagnetic torque can be expressed as

$$T_e = p_0\left(M_{fm}i_f i_q - M_{cpm}i_{cp}i_d\right) \tag{3.63}$$

According to the above analysis and derivation, if combining formula (3.62), (3.63) and (3.56) with each other, the mathematical model of a four-phase air-core pulsed alternator under the rectangular axis coordinates can be obtained.

Through coordinate transformation, the mathematical model of the four-phase air-core pulsed alternator is greatly simplified: the voltage equation of the motor becomes the constant coefficient differential equation, and the electromagnetic torque is decoupled. These can theoretically be solved by analytic calculation. Even when using software such as MATLAB SIMULINK or SABER to build analog block diagrams for circuit simulation calculations, the complexity of the system model will be significantly reduced, which effectively improves the simulation calculation efficiency of the system.

With the mathematical model of the motor and the inductance formula derived in Sect. 3.4 of this chapter, the performance of the air-core pulsed alternator can be calculated quickly by solving the state equation or establishing the circuit simulation model, the connection between the variables can be established, and the guidance can be provided for the direction of optimal design of the motor in the early stage.

3.7 Finite Element Modeling Method of the Pulsed Alternators

Assumptions and simplifications are used in the calculation of inductance and the establishment of mathematical modeling, which leads to clear physical concepts but poor calculating accuracy, especially for the air-core pulsed alternator. The internal inductance and load are only at the micro-henry and milliohm level. The output performance of the motor is greatly affected by the motor parameters. Therefore, it is necessary to establish the simulation model of the air-core pulsed alternator by using the time-stepping finite element method with higher computational accuracy for the sake of more accurate simulation results.

The finite element model of the air-core pulsed alternator is established with software Maxwell. On the one hand, Maxwell enables the calculation of dynamic inductance and eddy current effects with rotor motion in transient simulation. On the other hand, there is a seamless surface in the temperature analysis modules among Maxwell and Simplorer (a circuit software) which enables the transmission of two-way time-stepping data. This greatly facilitates the establishment of joint simulation models and temperature field analysis.

For axial symmetric motors, the two-dimensional finite element model can generally meet the needs of calculation accuracy. However, for the air-core pulsed alternator, because there is no iron-core guidance, the magnetic field produced by the air-core armature or air-core excitation windings is featured with obvious three-dimensional characteristics in the spatial distribution. Therefore, the parameters and motor characteristics of the air-core pulsed alternators can be determined more precisely by a 3D model. Considering the long time taken by three-dimensional finite elements, this paper makes appropriate corrections to the two-dimensional model based on the results of 3D field calculations, introduces the concept of inductance parameters at the end, uses a two-dimensional model in qualitative sensitivity analysis and optimization design, and a three-dimensional model to make a quantitative determination.

In the establishment of the electromagnetic field simulation model of the air-core pulsed alternator, in order to shorten the calculation time, according to the symmetry of the motor along the circumference, the model can be reduced to ½p of the actual model simulation, as shown in Fig. 3.11. For the structural force elements made of composite materials, such as yokes of stators and rotors, and the banding

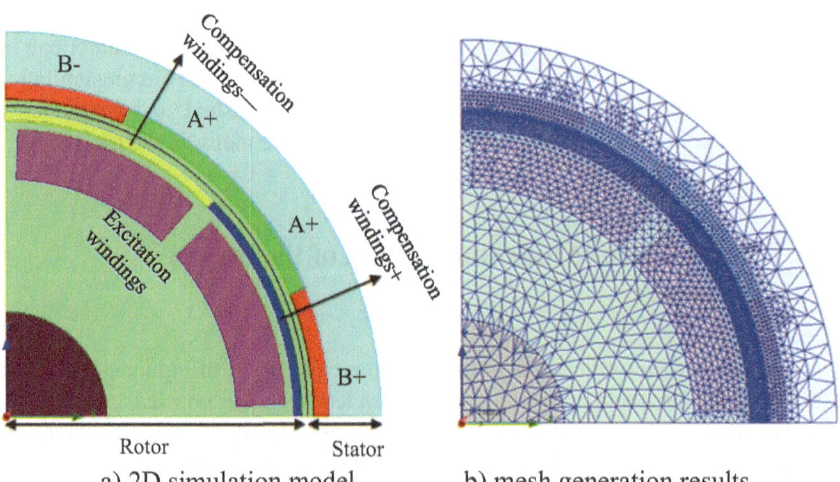

Fig. 3.11 2D simulation model and mesh generation results

of rotors, etc., because their magnetic conductivity is on the similar level of airs, they can be replaced by the air regions of stators and rotors. During simulation, the eddy current calculation options, such as electromagnetic shielding and conductive compensation cylinders (passive compensation structures only) are activated for the parts that produce the eddy current. The armature windings generally use shunt wound of Leeds lines and winding, so the eddy current effect is negligible. Each winding in the motor is in the form of an external circuit and activates the connection option of software Simplorer.

Maxwell transient analysis does not support mesh adaptation, therefore manual sectioning is the only option. For each new step of the rotor, the software meshes the air gap again to accommodate the new position of the rotor. Therefore, the degree of density of the mesh directly affects the velocity and accuracy of simulation calculation. Because it's faster to use 2D model for calculation, the grid can be properly encrypted to make the solution as accurate as possible.

For 3D models, the polygon surface is used to replace the air area and the cylindrical face of the winding, which reduces the curvature of the model and can obtain better mesh generation results. The preliminary process of mesh generation can be run in the static magnetic field solver first with the mesh from the self-adaption of the software as the preliminary mesh generation result. It will be imported into the transient solver. Then according to experience a step-by-step refinement of the mesh will be carried out from coarse to fine. When there is no significant change in the refinement results, it is considered that the final results can meet the requirements of accuracy. The 3D air-core pulsed alternator model and the mesh generation results are shown in Fig. 3.12.

When sectioning, special encryption is required for the windings, air gaps, shield, and the compensation shield since Maxwell 3D only supports the tetrahedral meshes. For the compensation shield with a small value of radial thickness and a big value

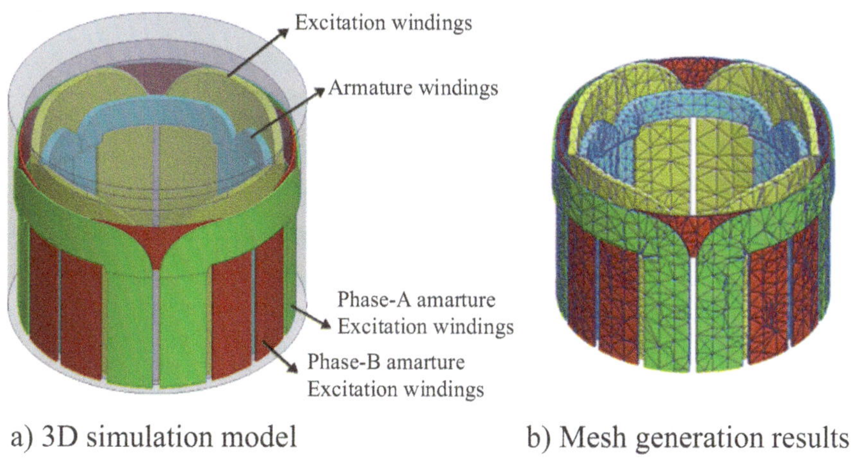

Fig. 3.12 3D simulation model and mesh generation results

of axial length, the division strategy of multi-layering of the solid model can be adopted. That is to say, the single layer compensation shield is composed of multi-layered structures that come into contact with each other, so as to obtain the result of layered encryption and relatively sparse mesh generation results in the axial direction, which improves the efficiency of simulation calculation.

Simulating the discharge characteristics of pulsed alternators also requires the establishment of corresponding load models. Detailed analysis and modeling methods will be shown in Chap. 7 of this book.

3.7.1 Design Process of Pulsed Alternators

Pulsed alternators are mainly used in the field of electromagnetic emission. Different from conventional alternators, pulsed alternators generally do not involve the concept of rated power, its output power is directly related to the load, so in the design of the motor, the requirements for energy and power of the load need to be considered. The energy demand of the load determines the main size and energy storage of the pulsed alternator, and the power demand of the load determines the output voltage and current of the pulsed alternator. The basic steps in the electromagnetic design of the pulsed alternator are shown in Fig. 3.13.

1. According to the electromagnetic transmitter's kinetic energy at outlet and the performance of continuous occurrence, the inertial energy storage of the rotors of pulsed alternators can be determined, and the velocity and main size of the motor.
2. According to the length and velocity at the outlet of the electromagnetic transmitter, the acceleration time of the emitter, that is, the pulse width of the output current can be calculated, and whether a single-phase structure or a multi-phase

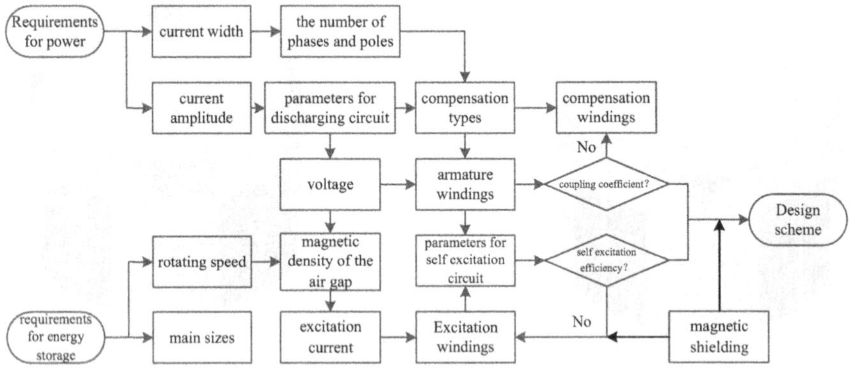

Fig. 3.13 Design flowchart of the pulsed alternators

structure is used in the pulsed alternator can be determined, and then analyse numbers of pairs of the pole and other major topology of the motor.

3. According to the acceleration and the inductance gradient required by the emitter, the output current I_L of the pulsed alternator is determined; According to the output voltage and the parameters of resistance, the output voltage and discharge circuit parameters of the pulsed alternator, in particular the equivalent inductance of the armature winding can be drew. Targeting at equivalent inductance and coupling coefficients, not only the compensation forms, but also the discharge circuit parameters of the nuclear power windings and compensation windings is expected. In this way, the armature windings and the compensation winding parameters can be further calculated.
4. According to the output voltage, armature winding parameters and speed determine the magnetic density of the air gap in the pulsed alternators, and then the calculation of the excitation current and excitation winding parameters in on the table. Taking required self-excitation efficiency as the goal, the self-excitation circuit parameter composed of the armature winding and excitation winding should be checked for further investigation of armature winding and excitation winding parameters.
5. According to the requirements of the volume and magnetic field shielding of the pulsed alternator, the shield is designed for the motor; and according to the size of the reaction effect of the shield to the motor, the excitation current and excitation winding parameters are further carried out to maintain the output capacity of the motor unchanged.
6. Through the simulation software, the motor is simulated and the motor's no-load and discharging characteristics are analyzed to meet the demand. If not, readjust the motor design parameters.

3.8 Case Studies of the Design

3.8.1 Proposal for Double-Axis Compensation

On the basis of passive compensation structure, this book introduces the design and analysis process of pulsed alternators by taking a double-axis compensation structure as an example. The pulsed alternator of double-axis compensation structure refers to the orthographical arrangement of motor's compensation winding and the electrical angle of excitation winding. During discharge, the compensation winding provides the cross-axis compensation, and the excitation winding in the state of continuous flow short-circuit is used to provide direct-axis compensation for the armature winding. The diagram of a four-pole rotating excitation double-axis compensation air-core CPA is shown in Fig. 3.14.

Double-axis compensation is a special passive alternation structure that integrates the advantages of passive compensation, through the combination of cross-axis compensation and direct-axis compensation, no matter what position the rotor is

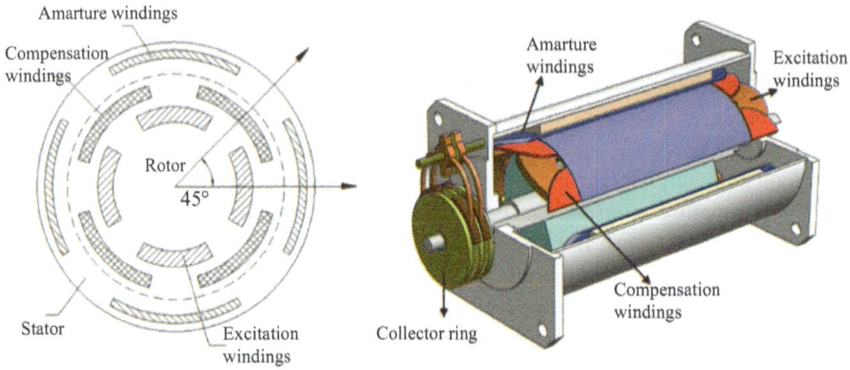

Fig. 3.14 The diagram of a 4-pole rotating excitation double-axis compensation air-core CPA

in, the armature winding can get a more uniform compensation to maintain a lower inductance of the motor. At the same time, this orthogonal arrangement decoupled the magnetic circuit coupling between the compensation winding and the excitation winding. The compensation winding is not inducted during self-excitation, and its time and system efficiency are greatly improved compared to the passive compensation structure.

When the CPA discharges to the load, there are three circuits in the motor, namely armature, the cross-axis compensation and the direct-axis compensation (excitation). Ignoring the resistance in the circuit, the voltage equation in each circuit is:

$$\begin{cases} U_a = L_a \frac{di_a}{dt} + M_{ac} \frac{di_c}{dt} + M_{af} \frac{di_f}{dt} \\ 0 = M_{ac} \frac{di_a}{dt} + L_c \frac{di_c}{dt} \\ 0 = M_{af} \frac{di_a}{dt} + L_f \frac{di_f}{dt} \end{cases} \quad (3.64)$$

Subscript a, c, and f represent armatures, compensation, and excitation windings respectively. Mutual inductance is related to the rotor position:

$$\begin{cases} M_{ac} = M_{acm} \sin \theta \\ M_{af} = M_{afm} \cos \theta \end{cases} \quad (3.65)$$

θ the electrical angle between the excitation winding and the armature winding.

Finishing (3.64),

$$U_a = L_a \left(1 - \frac{M_{ac}^2}{L_a L_c} - \frac{M_{af}^2}{L_a L_f} \right) \frac{di_a}{dt} \quad (3.66)$$

Taking into account the definition of coupling coefficient and the coupling coefficient expression derived from the formula, the equivalent inductance L_{aef} of the available armature winding is

$$L_{aef} = L_a\left(1 - k_{ac}^2 \sin^2\theta - k_{af}^2 \cos^2\theta\right)$$
$$= L_a\left[1 - \left(\frac{r_c}{r_a}\right)^{2p_0} \sin^2\theta - \left(\frac{r_f}{r_a}\right)^{2p_0} \cos^2\theta\right] \quad (3.67)$$

k_{ac} coupling coefficient between armature windings and compensation windings;
k_{af} the coupling coefficient between armature windings and excitation windings;
p_0 the number of the pairs of poles;
r the average radius of the position where the winding is located.

The above equation shows that when the equivalent inductance of the armature winding (L_{aef}) is non-related to the rotor position, the inductive compression is $(1 - k^2)$ times when uncompensated. The compensation effect is the same as the passive compensation, and the conductive compensation shield is replaced by two sets of orthogonal windings. When the coupling coefficients are not equal, taking the radius ratio of $r_c/r_a = 0.91$, $r_f/r_a = 0.85$ as an example, using the normalized algorithm, it is considered that when no compensation is made, $L_a = 1$, and the changing curve for calculating the equivalent inductance is shown in Fig. 3.15.

The results show that when the compensation winding and excitation winding are not on the same circumference, the equivalent inductance of the motor after the double-axis compensation fluctuates between the maximum compensation of the direct-axis and the maximum compensation of the quadrature-axis, as shown in Fig. 3.15a, the armature winding can obtain a more ideal compression ratio, and the equivalent inductance of the armature is always maintained at a low value. At the same time, the fluctuation range of the equivalent inductance is related to the pair number of the poles of the motor. The smaller the pair number is, the better the compensation of the armature winding will be. As shown in Fig. 3.15b, when the pair number p_0 is greater than 3, the compression rate of the armature winding has been less than 50%, the compensation effect is significantly reduced. Taking into

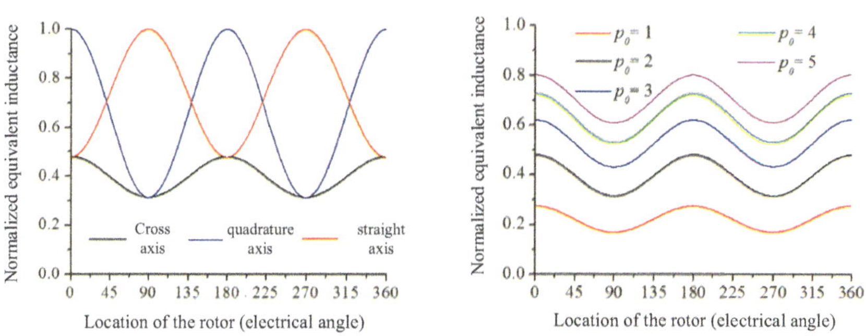

a) Equivalent inductance change curve ($p_0 = 2$)

b) Effect of pair numbers of poles on equivalent inductance

Fig. 3.15 How the equivalent inductance of double-axis compensation varies with the rotor

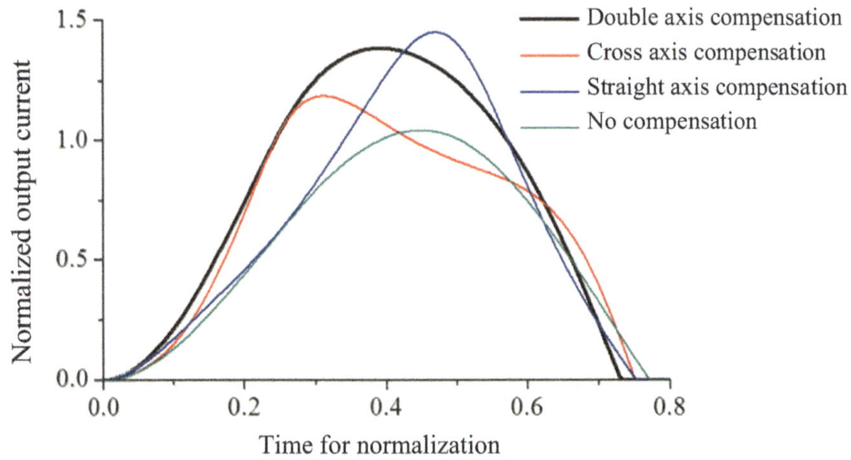

Fig. 3.16 Output current waveforms under different compensation structures

account the two-pole motor has a rotor shielding problem, the air-core CPA should be designed as a 4-pole or 6-pole structure.

Under the same load conditions, the finite element simulation calculation of the same reference CPA model is performed under the double-axis, quadrature-axis, direct-axis and no compensation, and the output current waveform is shown in Fig. 3.16.

The simulation results show that the CPA of the two-axis compensation structure is able to output the pulsed current of the sine-shaped structure, similar to the output current waveform of the passive compensation structure, and the effective value and rising rate of the output current are significantly higher than the uncompensated and direct-axis compensation structure, which shows that the double-axis compensation structure can make the motor have a constantly low inductance, improve the output capacity of the motor, and increase the flat-top ratio of the output current compared to the direct-axis compensation structure, which makes the load projectile obtain a more ideal average acceleration and peak acceleration ratio.

The double-axis compensation design also adds a certain flexibility that can be adjusted and optimized for CPA, which is matched between the two axes, making it easier for the CPA to obtain the ideal flat-top pulsed current waveform and improve the efficiency of the electromagnetic transmitter.

3.8.2 Equivalent Inductance Analysis of Double-Axis Compensation Air-Core CPA

Under the assumption that the winding current is the line current in sine distribution, the coupling coefficient between the armature winding and the compensation winding

k_{ac} is only related to the average radius ratio of the winding, not to the circumference ratio of the lower part of the compensation winding. In fact, the winding current within CPA is the distributed current on the plane, so it is necessary to use the finite element method to reverse out the inductance and coupling coefficient from the view of the electromagnetic field, and to examine its relationship with the ratio of the compensation winding.

With the 3D finite element model, the self and mutual inductance of three sets of windings in different positions of the rotor are calculated, and the relationship between the normalized equivalent inductance of the armature windings and the phase-belt of the compensation windings δ are calculated, as shown in Fig. 3.17.

In Fig. 3.17, the top line is the self-inductance of the armature when uncompensated. After normalizing it becomes "1", and the red dashed line of the sine change is the equivalent inductance of the armature when $\delta = 30°$ under the compensation of the direct-axis, which reaches the maximum value without compensation at the quadrature-axis position and the minimum value in the direct-axis compensation position. The other 4 curves are the equivalent inductances of the armature under the joint action of double-axis compensation at different angles, and it can be seen that the proportion of compensation windings and excitation windings directly affects the compensation distribution between the direct-axis and the quadrature-axis. When δ gets smaller, the ratio of the compensation winding conductor becomes smaller, the proportion of excitation winding conductor becomes larger, resulting in that the direct-axis compensation ratio is more sufficient than the cross-axis compensation; When $\delta = 45°$, the ratio of compensation windings and the excitations are equivalent, and the armature windings are compensated in a uniform way on the rectangular axis,

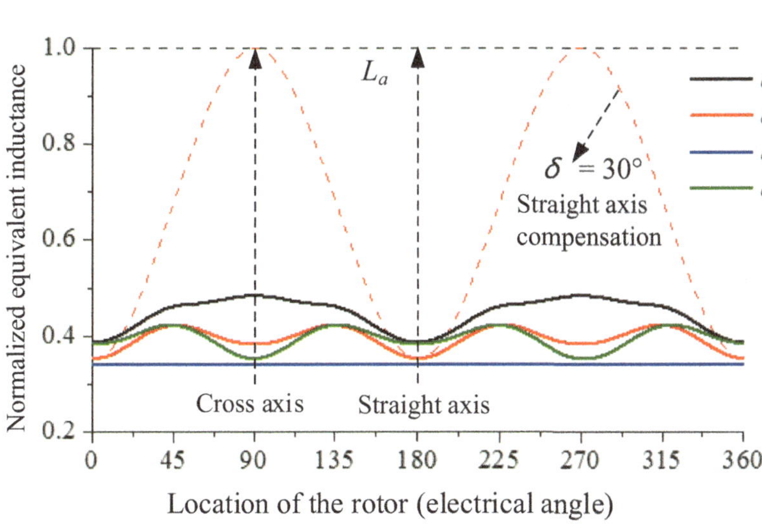

Fig. 3.17 Impact of δ on the armature equivalent inductance

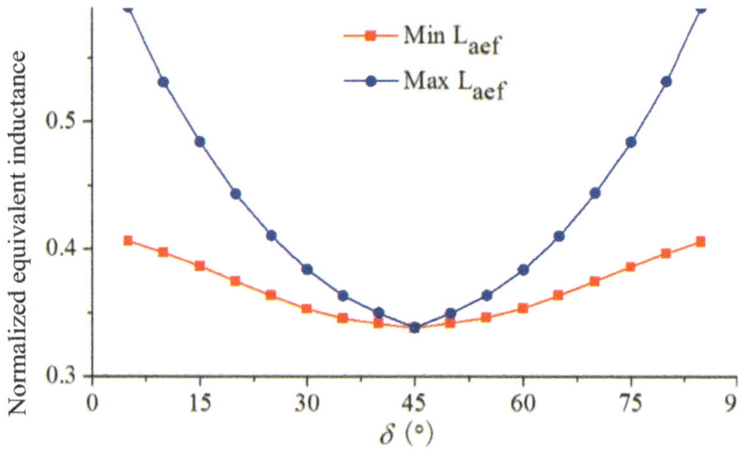

Fig. 3.18 Impact of the maximum and minimum value of the angle δ on the equivalent inductance of the armature

which equals to a compensation shield of the passive compensation structure. Therefore, the equivalent inductance can be maintained at a constant low value during discharge.

When the rectangular compensation is uneven, the equivalent inductance of the armature winding fluctuates with the rotor position between the maximum compensation of the direct-axis and the maximum compensation of the quadrature-axis, and the maximum and minimum value of the equivalent inductance changes with the angle δ as shown in the curve in Fig. 3.18. It can be seen that the larger the difference between the ratios of compensation windings and excitation windings is, that is, the more uneven the distribution of the rectangular axis compensation is, and the greater the maximum and minimum value of the equivalent inductance L_{aef} of the armature winding is, the larger the range of fluctuations in the inductive system will be. In order to make L_{aef} as small as possible, under the conditions allowed by structural design and the waveform of the air gap of the magnetic field, the difference between the ratio of the compensation windings and excitation windings of the double-axis compensation CPA should be as small as possible. The ideal value of the angle δ from Fig. 3.18 is 30–60°.

3.8.3 The Matching Design of the Double-Axis Compensation Air-Core CPA

In the two-axis compensation CPA of excitation windings and compensation windings with circumference design, the proportion relationship between the two windings needs to be considered in terms of the structural force, winding temperature rise and spatial position in addition to factors such as waveforms of the air gap of the magnetic field and rectangular axis compensation matching.

In terms of structural force, because during the instantaneous discharge the rectangular compensation current of compensation winding and excitation winding are obtained through electromagnetic induction, the induction current is always of opposite direction to the armature winding current. The armature magnetic field is compressed between the armature winding and the compensation element, so the component of the magnetic field of air gap is mainly tangential. According to the electromagnetic force $\boldsymbol{J} \times \boldsymbol{B}$, compensation windings and excitation windings during discharge are always subject to electromagnetic forces inside the rotor, as shown in Fig. 3.19.

The electromagnetic force subject to the winding is proportional to the current flowing through the winding. After the motor discharge parameters are determined, the induction currents of the compensation winding and excitation winding are mainly determined by the number of turns of the two sets of windings. Because the electro-resistance of the air-core CPA compensation winding and excitation winding is greater than the resistance by an order of magnitude, and the circuit is mainly inductive, basically the induction current of the two sets of windings is inversely proportional to the number of windings, as shown in Fig. 3.20.

In the double-axis compensation CPA, the excitation winding requires not only the direct axis compensation, but also the main pole magnetic field for CPA, so when the number of turns of excitation winding double or halve, in order to maintain the same air gap magnetic field and output voltage, the excitation current also needs to be halved or doubled. In this case, when the numbers of turns are different, the initial currents of the excitation winding group are not the same at the beginning

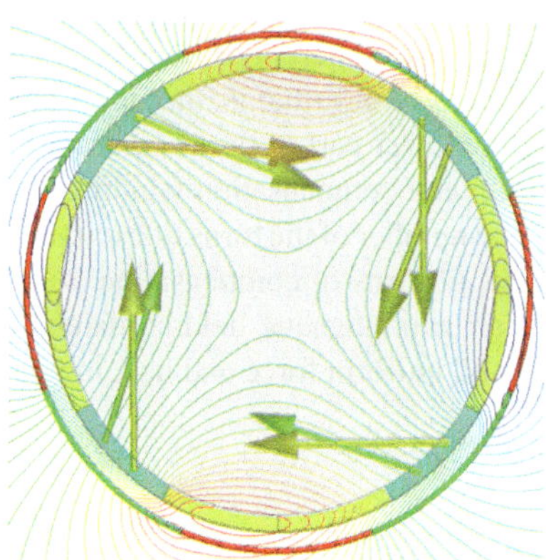

a) Compensation winding force conditions under quadrature-axis compensation

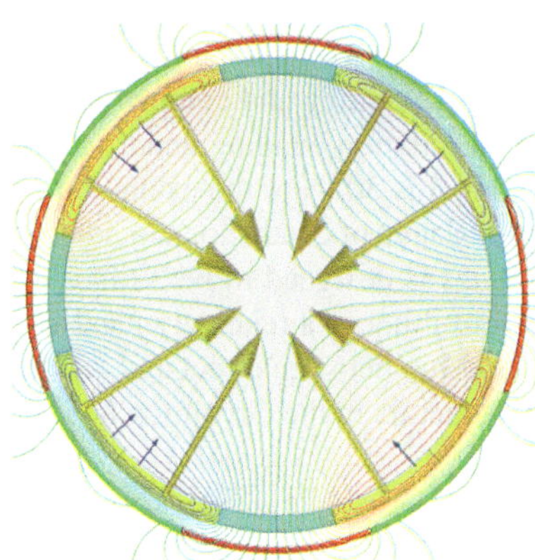

b) Excitation winding force conditions under direct axis compensation

Fig. 3.19 The electromagnetic force of compensation and armature windings during discharge

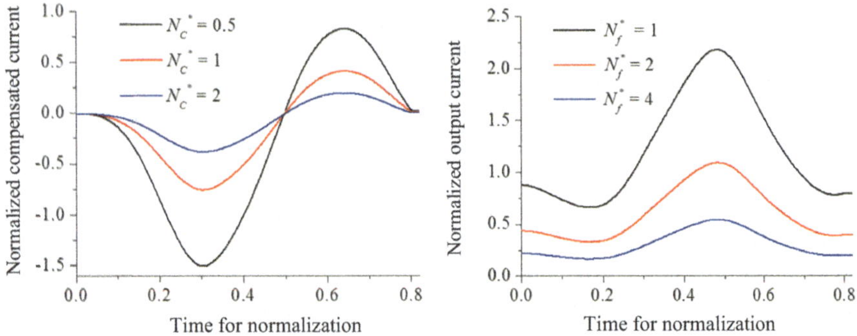

a) Compensation winding induction current b) excitation winding induction current

Fig. 3.20 Effect of winding turns on induced current

of discharge. However, from the range of current fluctuations, basically the induced current is still inversely proportional to the number of turns.

Taking into account the superposition of the initial currents before the excitation winding discharge, as shown in Fig. 3.20, when the number of turns for compensation windings is "1" and the number of turns for excitation windings is "2", that is, when the number of turns for excitation windings is twice the number of the number of turns for compensation windings, the peak and effective values of the two sets of winding induced currents are similar. If the same winding area and conductor material are used, the force area of the two windings is similar to the heat rate. On the one hand, the force and the heat are uniform in circular direction, on the other hand, the same process structure and cooling method can be used.

Integrating the three aspects discussed in this section, for the same circular design of the double-axis compensation air-core CPA, the circumferential ratio between excitation winding and compensation winding should be 2:1, that is, the lower part of the excitation winding under one pole is 60°, the lower part of the compensation winding δ is 30°. For the excitation winding group and compensation winding, the sectional area of conductor should be the same. The number of the turns of excitation windings is designed to be 2 times as the number of turns of compensation windings, resulting in a relatively ideal waveform of air gap magnetic field, a matching relationship between direct and quadrature-axis, and uniform force and heat ingress of conductors in the circumference.

Table 3.2 Design parameters for double-axis compensation air-core CPA

Mechanical parameters		Electromagnetic parameters	
Rated speed (rpm)	10,000	Rated excitation current (kA)	45
moment of inertia J (kg m^2)	27.3	Peak value of no-load voltage (kV)	3.5
Rotor length (m)	0.95	Numbers of turns of armature windings (parallel)	4
Rotor outer diameter (m)	0.6	Number of turns of compensation windings (parallel)	6
Coil effective length L_e (m)	0.8	Number of turns of excitation windings (series)	12

3.8.4 Design Parameters and Simulation Model of Double-Axis Compensation Air-Core CPA

In order to compare and explain the application of the two-axis compensation structure in the single-phase and multi-phase CPA system, this paper selects cannon-caliber orbital guns and medium-sized orbital guns as CPA loads, referring to the two-phase double-axis compensation air-core CPA with small pulse width and a canon station of storage of 15 MJ. Its main design parameters are shown in Table 3.2.

With the help of the 3D electromagnetic field joint simulation model calculation system, this section mainly investigates the magnetic field distribution and output capacity of CPA discharge moment. In order to improve the speed of simulation calculation, the rapid joint simulation model is adapted here, as shown in Fig. 3.21, with an ideal capacitance discharge instead of self-excitation magnetic process. The initial excitation current is provided directly to the CPA, saving the time required by the simulation of the self-exciting process.

Excitation capacitor C_f provides an initial excitation current of 45 kA after cutting out the excitation circuit. There is a short circuit of excitation winding after flowing through the fly-wheel diode $V_{D.}$ The phase A and B of armature winding are connected to the loaded orbital gun model by the discharge rectifier FR_a and FR_b in the corresponding trigger angle, and discharge to the orbital gun through the control signal Sgn_a and $Sgn_{b.}$

3.8.5 Analysis of Single Pulse Discharging Characteristics of Double-Axis Compensation Air-Core CPA

This section takes Cannon-caliber orbital guns as a load to examine the cross and direct-axis compensation and output capacity of single-phase double-axis compensation air-core CPA. A typical Cannon-caliber orbital gun is selected as the load,

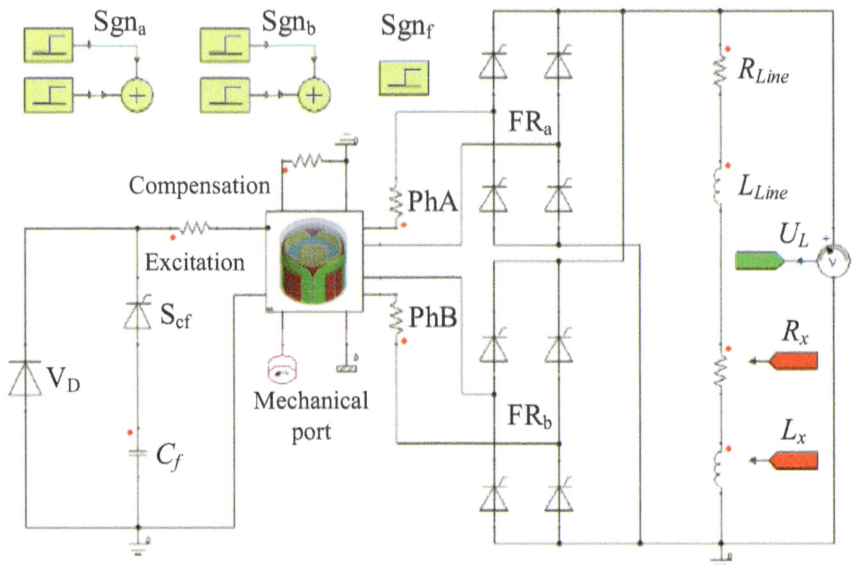

Fig. 3.21 Joint simulation model of capacitive excitation rapid air-core CPA

with a length of 2 m, a resistance gradient of $R' = 0.1$ mΩ/mm, an inductive gradient $L' = 0.5$ μH/m, a projectile mass of 100 g and a target launch speed of 2 km/s. In the discharging circuit, the self-inductance of the armature winding is 4.84 μH; the self-inductance of CPA after two-axis compensation CPA is 1.78 μH; the internal resistance of the armature winding is 0.5 mΩ; the resistance of the switch and the circuit is 0.5 mΩ; and the inductance is 1 μH.

(1) Comparison between the discharging characteristics of double-axis compensation and direct-axis compensation

In the simulation of single-phase single pulse discharge, under the compensations of double-axis and direct-axis, the output capacity of the air-core CPA when only considering phase A of the armature winding is shown in Fig. 3.22. For the convenience of comparison, the falling process is neglected, and the speed of CPA is assumed to be 10,000 rpm constantly.

As can be seen from the comparison of output current in Fig. 3.22a, the current waveform of the CPA approximates sine after the double-axis compensation, which is consistent with the theoretical analysis. The current waveform is ideal for the single-phase electromagnetic orbital gun system, compared with the direct-axis compensation. The effective value and peak ratio of current is increased from 0.52 to 0.65, which makes the projectiles accelerate smoothly in the orbital gun, reduces the instantaneous stress impact on the orbital gun, extends the service life of the rail gun, and also lowers the requirements for the discharging switch tube.

Although the output current peak has dropped from 992 to 907 kA, the effective value has been increased from 496 to 560 kA, there is still a greater average

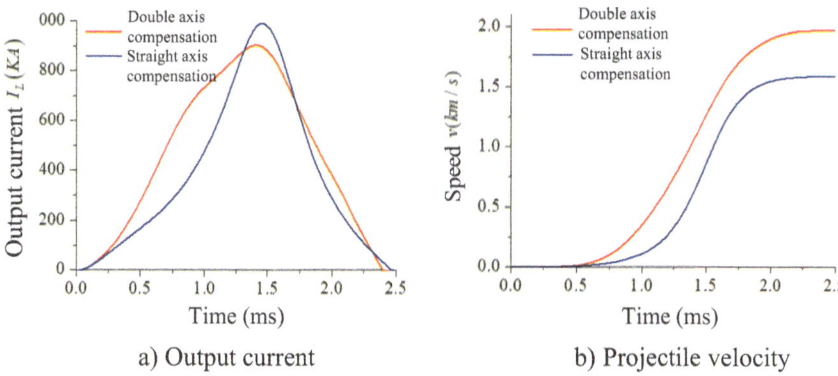

Fig. 3.22 Output capacity for double-axis compensation and direct-axis compensation

acceleration for the projectiles and an increase in the output speed of the projectiles from 1588 to 1974 m/s, as shown in Fig. 3.22b. The quadrature-axis compensation winding is designed for short circuit, no additional switching devices and cables need to be connected. In this regard, when the quadrature-axis compensation winding uses aluminum windings of a similar density with compound filler, the quality of the CPA system is basically the same, and the improvement of the dynamic energy of the orbital gun outlet increases the transmission energy density of the system by 54.5%.

When the two-axis compensates the CPA discharge, the compensation windings and excitation windings induce the current at the corresponding position respectively to provide the cross and direct-axis compensation for the armature windings. Because the compensation windings located on the quadrature-axis do not induce the current during self-excitation, the compensation current changes from zero when discharge, while the excitation windings need to provide a main pole magnetic field to the CPA so that they operate in a continuous short-circuit state. Therefore there is an initial current of 45 kA at the beginning of the discharge. When the trigger angle is 0° and the single-phase single pulse is discharging, the induction current of compensation windings and excitation winding currents are shown in Fig. 3.23.

When the compensation winding axis coincides with the discharging phase-A armature axis (in this case it's 0.75 ms), the electromagnetic coupling between the two sets of windings is the strongest, reaching the maximum value of the quadrature-axis compensation. Comparing the distribution of vector magnetism within CPA under double-axis compensation and direct-axis compensation, as shown in Fig. 3.24, it can be seen that because the compensation winding wire rotating to the phase-A armature winding conductor at this time, the compensation current changes the magnetic field distribution within the CPA, and the flux is compressed between the compensation winding and the phase-A armature winding group. The equivalent inductance of the armature winding is significantly reduced, so it is possible for the two-axis compensation CPA to gain a higher current rising rate from the beginning of discharge.

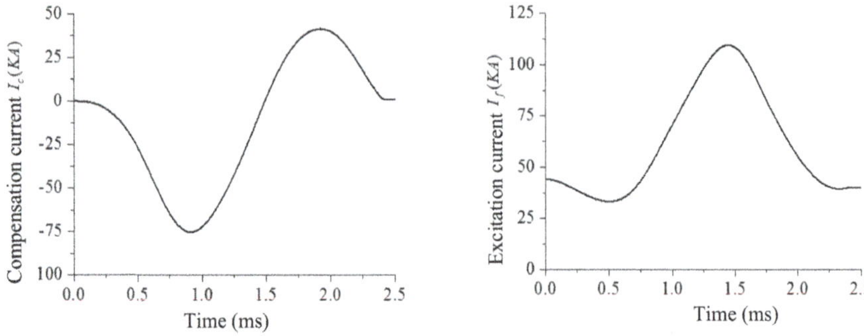

a) Induction current of compensation windings b) Induction current of excitation windings

Fig. 3.23 Induction current of compensation windings and excitation windings during instantaneous discharge

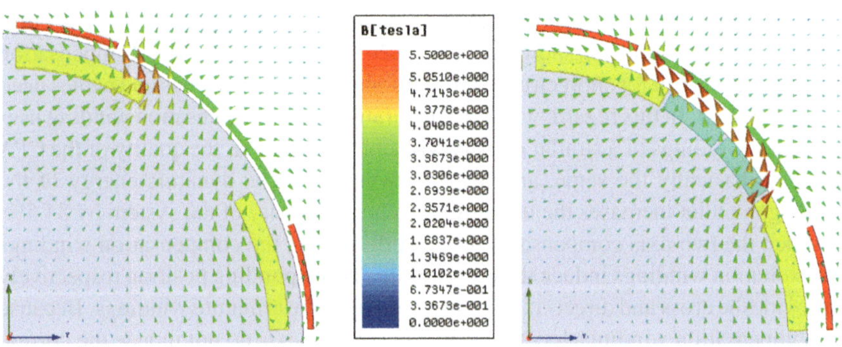

a) Direct-axis compensation b) Double-axis compensation

Fig. 3.24 The vector magnetic distribution of the air-core CPA under the maximum quadrature-axis compensation

When the excitation winding rotates to the position facing the phase-A armature winding (in this case is 1.5 ms), the excitation winding is in continuous short circuit and receives the maximum induction current, i.e. the direct-axis compensation reaches the maximum value, as shown in Fig. 3.25, the flux is compressed between the excitation winding and the phase-A armature winding, and also obtains a lower armature equivalent inductance. Since the armature winding output current is close to the maximum at this time, the magnetic density in the motor is larger than under the cross-axis compensation.

When the axis of the excitation winding and the compensation winding do not coincide with the armature winding axis (as shown in Fig. 3.26, 1.05 ms after discharging), the coupling of the compensation windings and the excitation windings and the armature windings exists, while providing compensation for the armature windings, so that the flux of the armature windings is always compressed in the air.

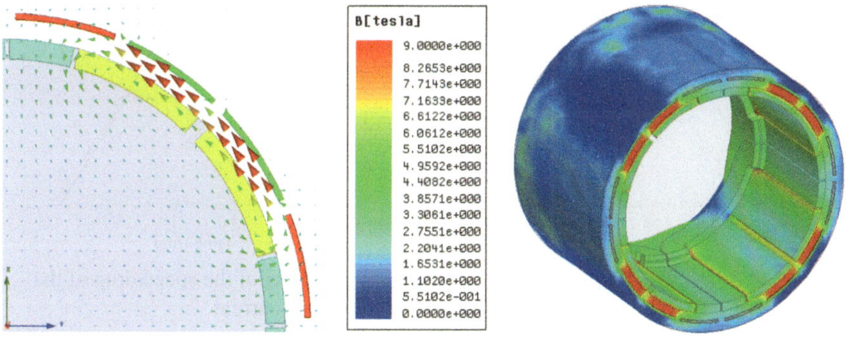

Fig. 3.25 Distribution of magnetic density of the air-core CPA when the direct-axis compensation is of the maximum value

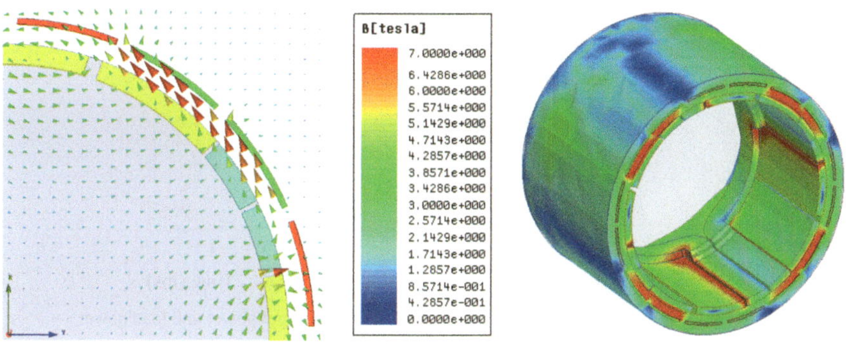

Fig. 3.26 Distribution of the magnetic density of air-core CPA under two-axis compensation when t = 1.05 ms

Thus, the double-axis compensation CPA obtains a constantly low internal inductance and verifies the validity of the two-axis compensation theory from the view of electromagnetic field.

(2) Output capacity of the double-axis compensated CPA

The above simulation is carried out under the assumption of the constant speed of CPA. In fact, in order to avoid the impact of the instantaneous impact on the prime mover, it should be separated from CPA before the discharging began. CPA rotates freely depending on the inertia. When considering the change of the speed, the mechanical port in the joint simulation model should use the Mass module, where given the inertia of 27.3 kg m^2 and the initial speed of 10,000 rpm. The output current and curve of the change in the speed of CPA at the time of inertial rotation is shown in Fig. 3.27.

As can be seen from Fig. 3.28a, because the speed drops during the instantaneous discharging, there's a decrease in CPA's induction voltage and discharge circuit's inductive reactance at the same time. In this way the output current only slightly

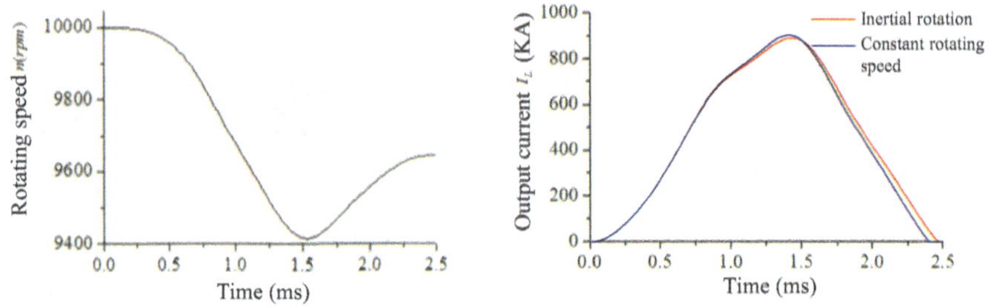

a) change in speed

b) Effect of the speed drop of the output current

Fig. 3.27 Changes in output current and speed during inertial rotation

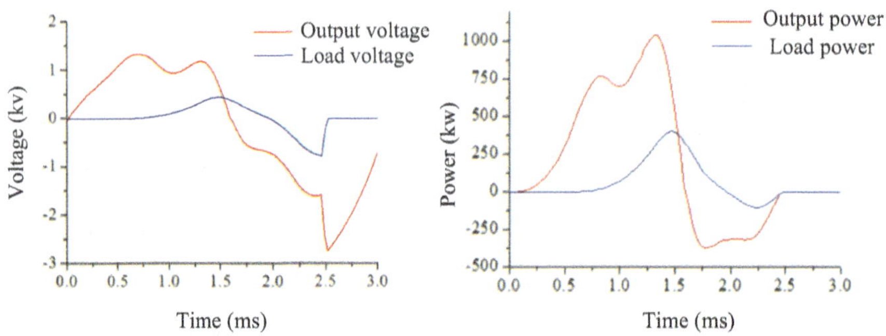

a) Output voltage and load voltage

b) Output power and load power

Fig. 3.28 The voltage and power of double-axis compensation CPA

decreased, and due to the reduction of motor frequency, the discharge pulse width slightly increased, but the overall discharge current waveform and constant speed simulation results are similar. Therefore, the abovementioned assumptions of simulation of constant speed are reasonable. It can be seen in Fig. 3.28b that at the beginning of the discharge, the speed of the motor is reduced, the inertial energy storage is converted into electric energy, and the momentary torque impact can reach 1.9 MN m. After 1.5 ms of discharge, the voltage becomes reverse, the motor output power is negative, the electric energy stored in the circuit is recovered, and the speed of the motor is recovered. After the discharge, the CPA's speed is reduced to 9646 rpm, and the inertial energy storage loss is $\Delta E_r = 1040$ kJ.

When considering the speed change, the voltage and power of the double-axis compensation air-core CPA are shown in Fig. 3.29. It can be seen that the internal resistance of the orbital gun is very small and changes from zero, the CPA output voltage and power mainly loss in the resistance of the line and switch. The output instantaneous peak power of CPA is up to 1040 MW, and the peak value of the input power on the load railgun is 400 MW.

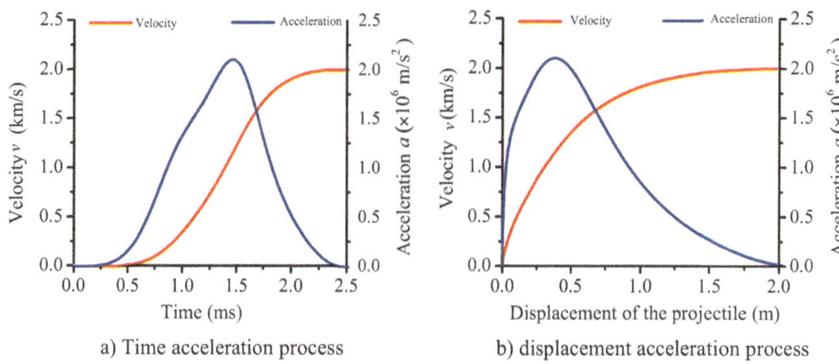

Fig. 3.29 The speed and acceleration of the projectiles in the orbital gun

Under the double-axis compensation CPA output current, the acceleration process of the projectile in the orbital gun is shown in Fig. 3.29a. The acceleration process lasts for 2.5 ms, the projectile obtains the outlet speed of 2010 m/s, and the maximum acceleration is 2.12×10^6 m/s^2, which meets the expected design target. From the displacement acceleration process in Fig. 3.29b, it can also be seen that the projectile accelerates at the exit position of the 2-m long orbital gun at a speed of 0. This suggests that the CPA output current is 0 at that position, which inhibits the production of the exporting arc of the orbital gun, eliminates the arc-killing mechanism of the orbital gun, and improves the launch efficiency of the orbital gun.

Based on the above simulation results, the launch energy of the orbital gun is $E_m = 202$ kJ, and the kinetic energy loss of the CPA ΔE_r is 1040 kJ, and the transmission efficiency η_r is 19.4%. It should be noted that this transferring efficiency only takes the kinetic energy loss of CPA during discharge into account. The kinetic energy loss caused by the loss of excitation circuit during the self-excitation process and energy recovery process will be analyzed in the next chapter.

(3) Effect of speed on output capacity of double-axis compensated CPA

CPA-driven orbital guns are generally in the operating mode of multiple continuous firings, which means, after the prime mover drives CPA to the rated speed, the CPA completes a continuous discharge of the orbital gun in a few seconds or tens of seconds, and the prime mover is re-connected to the CPA after one discharge, preparing for the next round of firing for the CPA's complementary kinetic energy.

In addition to the heat of the motor, the number of CPA's continuous occurrences in a round is also subject to the CPA's energy storage and speed constraints. When the motor speed reduces to a certain level, the CPA's self-excitation speed slows down until the excitation cannot even be carried out, so the CPA will not be able to convert all the stored energy into electrical energy like a capacitor. In general, the loss ratio of single-wheeled electric energy should be less than 0.5. Taking into account that the motor's energy storage and speed square is proportional, in order to maintain the output capacity in the continuous bursting process, CPA is still required to still be

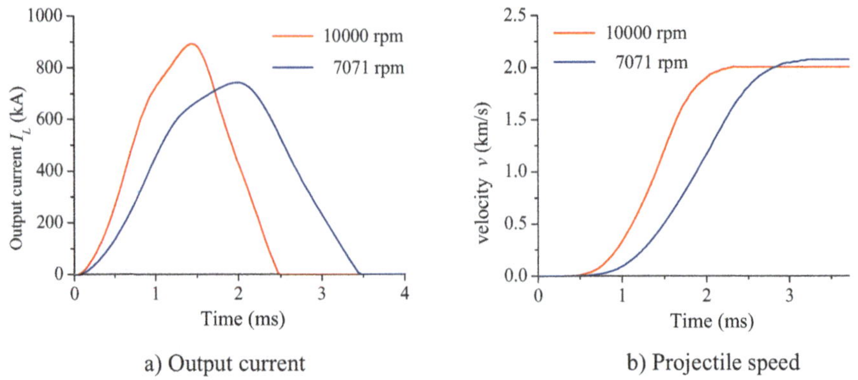

Fig. 3.30 Effect of speed on the output capacity of double-axis compensation air-core CPA

capable of driving the projectile to the target launch speed at a 70.71% rated speed. When the initial speed of the double-axis compensation CPA decreases to 7071 rpm, the output current waveform of the motor and the resulting projectile outlet speed are as shown in Fig. 3.30.

It can be seen that due to the reduction of motor speed, CPA's output current amplitude decreased from 895 to 745 kA, but the output pulse width increased from 2.5 to 3.5 ms. In general, the speed obtained by the projectile was increased from 2010 to 2080 m/s, indicating that CPA still has good output capacity at low speeds. This is because the CPA's discharge circuit is generally inductive, and the circuit resistance decreases as the speed gets lower. In this case, although the speed of the motor is reduced by 30%, the output current is reduced by only 17%, since the current duration is lengthened, projectile acquires a higher exit speed at a low rotational speed. That means the CPA decelerates and does not have a negative impact on the output capacity.

3.8.6 Multi-Pulse Discharge Characteristics Analysis of Double-Axis Compensation Air-Core CPA

In the future, the orbital gun will continue to develop in large-caliber, high initial speed, high outlet kinetic energy. The length of the orbital gun and the quality of the projectile will be significantly improved, which requires the CPA to be able to provide the orbital gun with a sufficiently wide pulse current. The CPA with multiphase structure, discharges to the load after the rectifier, and decouples the constraint relationship between motor speed and pulse width, for the sake of both sufficient pulse width, and a higher speed to a greater energy storage density. This defines the future development direction of CPA.

This section takes medium-diameter orbital guns as a load to examine the multi-phase multi-pulse output capacity of the double-axis compensation air-core CPA. As shown in Fig. 3.19, CPA adopts a two-phase structure, and the two sets of armature windings A and B, which are orthogonal in electrical angle. They discharge to the load through their respective full-bridge rectifier circuits separately. The load selects a medium-sized orbital gun with a length of 5 m, a projectile mass of 300 g, a target emission speed of 2 km/s, and other simulation conditions which are the same as single pulse discharge. The ideal capacitive excitation rapid simulation model is still in use.

Studies have shown that the structural strength of the orbital gun is mainly limited by the peak acceleration of the projectile, so for heavy load launches of more than 100 g, a higher ratio of average to peak acceleration value means longer acceleration times, resulting in higher gun velocity or shorter orbital lengths. Therefore, the input pulsed current waveform, which is approximately flat-top, is one of the necessary indicators for diameter orbital guns of medium size or above.

The output current of multiphase CPA is obtained by multiple pulsed current modulation, because the rotational speed of the instant discharge of CPA and excitation current are beyond control. CPA can generally only adjust the magnitude and pulse width of each pulsed current by controlling the trigger angle of each pulsed current, and thus obtain the ideal flat-top current waveform of the orbital gun.

According to the circuit theory, the CPA-driven railgun system can be regarded as an RL series circuit. Discharging is approximately corresponding to the zero state of sine voltage excitation. The equation for the turned-on circuit is

$$L\frac{di_L}{dt} + Ri_L = U_m \sin(\omega t + \varphi_u) \tag{3.68}$$

To solve the differential equation, the discharge current can be expressed as

$$i_L(t) = \frac{U_m \sin(\omega t + \varphi_u - \varphi)}{\sqrt{(\omega L)^2 + R^2}} - \frac{U_m \sin(\varphi_u - \varphi)}{\sqrt{(\omega L)^2 + R^2}} e^{-\frac{R}{L}t} \tag{3.69}$$

where L and R are the inductance and resistance of the discharge circuit, Um is the armature voltage amplitude, and φ_u is the initial phase angle, which is also the discharge trigger angle, and $\varphi = arctan(\omega L/R)$.

The preceding expression of the formula (3.69) current is the forced component, which changes with the sine pattern of the same frequency as the outcast sine excitation. While the latter is the free component in exponential form, which tends to zero over time. The discharging trigger angle φ_u determines the size of the free component, so it has a straight effect on the magnitude and pulse width of the output current.

When the switch is closed, if $\varphi_u = \varphi$, there is a free component of the following item is zero, the circuit does not occur in the transition process and comes immediately into a stable state. Because the general discharge circuit is inductive, the trigger angle to meet the condition is close to 90°; if $\varphi_u = \varphi - \pi/2$, the latter free component is of the maximum value and the longest transition process. In addition, the largest instant value of the current is the closest to the stable one.

For the two-axis compensation CPA for the study, the total resistance of the discharge circuit at 10,000 rpm is $R = 1.1\ \text{m}\Omega$, the total electrical resistance is $\omega L = 6.87\ \text{m}\Omega$. After calculation, it can be figured out that $\varphi = 80°$, which means that when $\varphi = 80°$, the output current comes to a stable state immediately. This will facilitate the multiphase multi-pulse output to get a flat-top current waveform more easily. When the trigger angle φ_u is $-10°$, the output current pulse width and amplitude is the largest, since the thyristor can only come to be conducting in the positive voltage, the actual trigger angle of the maximum free component is 0°. This is favorable for a single-phase single pulse discharge to obtain a higher amplitude and a wider pulse.

Take the single phase-A winding's continuous output of three pulses as an example. At the trigger angle of 0 and 80°, the output current results under the effect of double-axis compensation and direct-axis compensation are shown in Fig. 3.29.

The results of the simulation are consistent with the theoretical analysis. It can be figured that the output current directly enters into the steady-state process when $\varphi_u = 80°$, and the output current frequency is the same as that of the motor. Under the double-axis compensation effect, the waveform of the output current is closer to sinusoidal, the ratio of effective value to peak value is increased from 0.65 to 0.69, and the current rises at a faster rate. If there is only the direct-axis compensation effect, the ratio of effective value to peak value of the current is increased from 0.52 to 0.61, and the output current is changed from spire to approximately triangular wave.

In addition, because the excitation winding works in the continuous flowing short-circuit state, the excitation current gradually decays during discharge, while the resistance of the orbital gun gradually increases with the projectile displacement. Therefore, the magnitude of the current gradually becomes smaller during continuous output. In order to help the orbital gun to obtain the ideal trapezoidal input current waveform, the CPA of the multi-pulse output shall modulate the trigger angle of the remaining pulses according to the magnitude of the last pulsed current.

The two-phase CPA outputs to the load by the output rectifier FR_a and FR_b, and there is a converting process between phases, so the oscillating process of the output current is slightly different from the single-phase output, except for the first pulse. Only when the anti-electric power of one phase armature is higher than the other one, the phase can be triggered. Ideally, the natural phase changing point should be 45°, but actually because the circuit is inductive, there should be a corresponding delay in the natural phase changing point. Hereby trigger angle can only be enlarged when it is greater than the natural phase changing point, that is, the output current can only be adjusted to be smaller.

Aiming at the ideal trapezoidal current waveform of the orbital gun, the pulsed trigger angle of the air-core CPA is modulated under the action of double-axis compensation and direct-axis compensation respectively, and the output current waveform of the motor is as shown in Fig. 3.30.

It can be seen that in the two compensation structures, after the modulation of 6 consecutive pulses, the two-phase air-core CPA outputs the ideal trapezoidal current

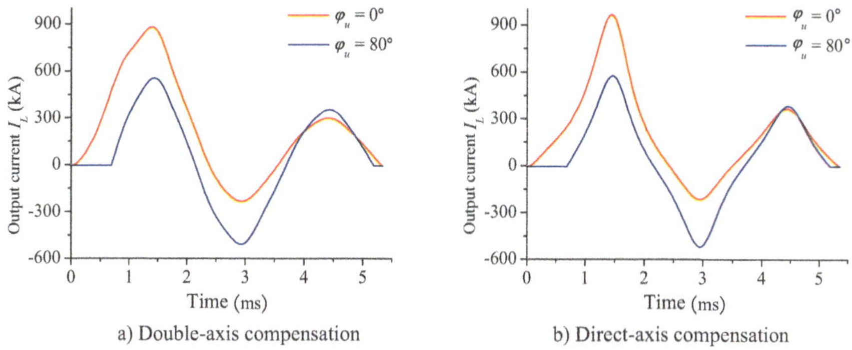

Fig. 3.31 Effect of trigger angle on output current

waveform, with the first trigger angle of the phase A and B being 72 and 86.4° respectively, and There is also a 14.4° of B since after the initial discharging of phase A, the excitation winding provides a direct-axis compensation effect. In order to keep up the output current amplitude with phase A, it's necessary for phase B to enlarge the trigger angle thus decrease the output current. When double-axis compensation CPA is carrying out multi-pulse outputs, compensation windings and excitation windings also provide compensation for the armature windings at the quadrature-axis and direct-axis positions. The current implied in the two windings are as shown in Fig. 3.31.

In pulsed modulation, in order to make the last pulsed output current the largest in amplitude, the trigger angles of the last pulse of the two phases under the two compensation structures are both natural phase-changing points. The difference between the two compensation structures is mainly reflected in the triggering moment of the two pulses in the middle. For the double-axis compensation structure, trigger angles of the two phases A and B are 18° and appear after the natural phase changing point. For the direct-axis compensation structure, the trigger angles of the two middle pulses are 10.8° and 7.2°, visible after the natural phase changing point. Such simulation shows that in order to obtain the flat top current, the double-axis compensation structure makes more use of the trigger angle to suppress the output current, at the expense of the capacity of getting a higher amplitude.

Under the action of the output current of the two compensation structures as shown in Fig. 3.30, the velocity and acceleration obtained by the orbital projectiles are shown in Fig. 3.32. The double-axis compensation CPA has an outlet velocity speed of 1890 m/s, and a peak ratio of 0.607 for the acceleration of projectile acceleration. The direct-axis compensation has an outlet velocity speed of 2020 m/s with a ratio of acceleration average to the peak value of 0.598. It can be objected that after pulse modulation, the two compensation structures have the ideal smooth acceleration. The direct-axis compensation CPA's projectile outlet velocity meets the requirements of the launch indicator, while this is not for the double-axis compensation CPA. Without imposing influences on the acceleration ratio, a larger output current can be obtained

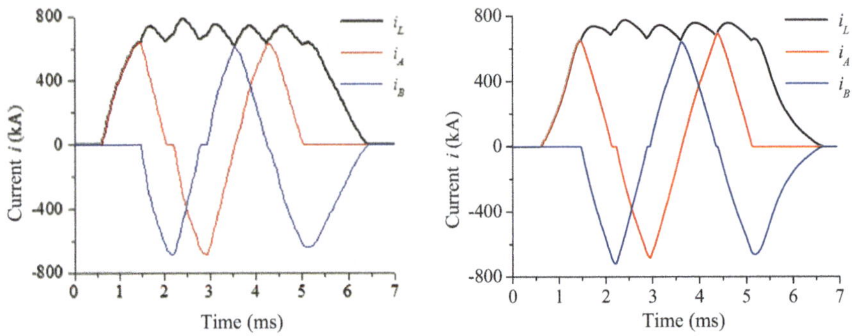

a) Double-axis compensation b) Direct-axis compensation

Fig. 3.32 Multi-pulse output of two-phase air-core CPA

by increasing the excitation current, so that the projectile sits to fulfill the export speed standard (Figs. 3.33, 3.34).

Through the above analysis, it can be seen that under the multi-pulse output of CPA, in order to ensure that each pulse is of the same amplitude, the first few pulses need to reduce the peak of the output current by increasing the trigger angle, so it's challenging to show the advantages of low inductance and high efficiency of double-axis compensation. On the other hand, when two-phase pulsed waveform overlays modulation, the approximately sine waveform of double-axis compensation has no advantage over the approximately triangular waveform of direct-axis compensation, but the triangular wave is more likely to achieve a superposition of the output current waveform of the flat top. The advantages of double-axis compensation in multi-pulse output are mainly reflected in the beginning and end of discharge. The rise and fall of the slope of the current is greater, which is conducive for a longer projectile acceleration time at the same pulse width and can slightly increase the ratio of average acceleration to the peak value of the projectile.

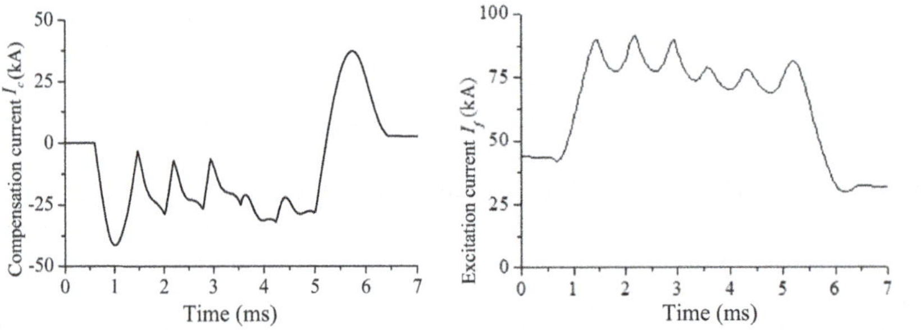

a) Induction current of compensation windings b) Induction current of excitation windings

Fig. 3.33 Induction current of compensation windings and excitation windings under multi-pulse discharging

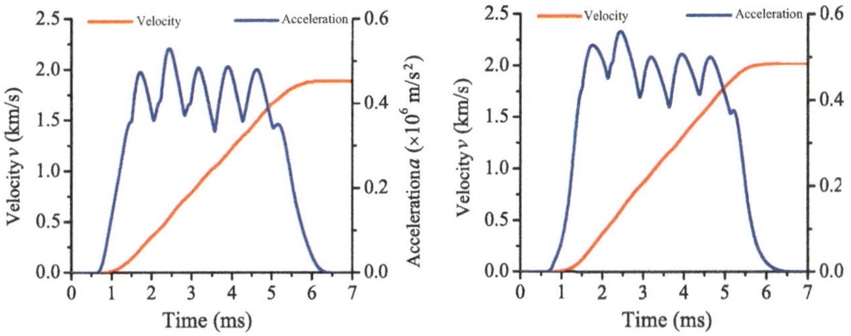

a) Double-axis compensation

b) Direct-axis compensation

Fig. 3.34 Speed and acceleration of projectiles under multi-pulse action

In summary, when the double-axis compensation CPA is driving orbital gun load under multi-phase and multi-pulse, it outputs the ideal trapezoidal current waveform. However, compared with the direct-axis compensation structure it does not have a clear advantage, taking into account that the additional compensation winding brings a certain degree of difficulty to the motor's structural design and process. This section proposes that the double-axis compensation structure is not suitable for multi-phase CPA-driven medium-length orbital gun system, and is more favorable for application in single-phase CPA-driven short-railgun system.

Chapter 4
Thermal Management of Pulsed Alternators

In order to resist the momentary discharge impact and high-speed centrifugal force, the stator and rotor winding of the air-core CPA is usually encased in composite materials and epoxy resins, and these materials have a very low thermal conductivity. This is a challenge for the air-core CPA winding to deliver an efficient transfer of the heat it generated to the outside world for convective cooling. In this case, the temperature of the winding increases gradually with the rising number of discharges, which limits the continuous operation capacity of the air-core CPA. Therefore, it is necessary to study on the temperature rise and cooling of the air-core CPA.

4.1 Temperature Field Analysis of Pulsed Alternators

4.1.1 Basic Heat Transfer Theories

In order to discuss the role of heat exchange in the motor of each component in the air-core CPA, this section briefly introduces the basic theory of heat transfer, which serves as the basis of heat transfer in the calculation of motor temperature field.

The concept of heat transfer in engineering is distinguished from the physical essence in terms of three basic forms, namely, thermal conduction, thermal convection and thermal radiation. The three forms and the basic calculation methods are introduced as following.

4.1.1.1 Thermal Conduction

When there is no relative displacement between the parts of an object, thermal transfer from the thermal motion of microscopic particles such as molecules, atoms, and free

S. Wu and S. Cui, *Pulsed Alternators Technologies and Application*,
https://doi.org/10.1007/978-981-33-4224-8_4

electrons is called thermal conduction. For example, the transfer of heat inside solid from the part of higher temperature to the part of lower temperature, and the transfer of heat to a lower-temperature solid from a higher-temperature solid with which it is in contact are both in this category.

By refining a large number of practical thermal problems, the law of thermal conductivity has been summed up as Fourier's law: the heat transmitted by unit time within an object is proportional to the temperature gradient within the object. It can be expressed as:

$$\vec{q} = -\lambda \frac{\partial t}{\partial x} \vec{n} \tag{4.1}$$

- q heat flux in W/m^2;
- λ thermal conductivity in W/(m·K). It represents the degree of thermal conductivity;
- t temperature in K;
- x coordinates on the thermal surface, in m;
- n The normal unit vector that passes through the isometric line of the point, pointing in the direction of the temperature increase.

4.1.1.2 Thermal Convection

Thermal convection refers to the transfer of heat arose from the relative displacement between the various parts of the fluid caused by the macro-motion of the fluid, and the heat transfer from the mixing of cold and hot fluids. In terms of the cause of the flow this concept can be divided into forced convection heat transfer and natural convection heat transfer. Forced convection heat transfer refers to the fluid is driven by external forces to produce the flow of heat transfer and natural convection heat transfer refers to the fluid due to its uneven temperature formed by the floating force when the spontaneous flow of heat transfer.

The basic calculation of thermal convection is Newton's law of cooling:

$$q = h\Delta t \tag{4.2}$$

- Δt temperature difference, which is always positive;
- h surface heat transfer coefficient (or thermal convection transfer coefficient) in (m^2·K) which represents the strength of heat transfer.

4.1.1.3 Thermal Radiation

Thermal radiation refers to the heat transfer process in which an object emits electromagnetic energy which is absorbed by other objects and transferred into heat. The calculation of the radiant heat flux of an actual object may take the empirical modification form of Stefan-Boltzmann Law:

$$\Phi = \varepsilon A \sigma T^4 \tag{4.3}$$

ε the emissivity of the object which is less than 1;
T thermodynamic temperature of the black body in K;
σ Stefan-Boltzmann constant, 5.67×10^{-8}W/(m^2·K^4);
A radiation of surface area, in m^2.

In motors, radiation heat transfer is only a small step of the whole operation process, especially for motor with forced convection cooling for which radiation is often ignored.

4.1.2 Calculation Method for Motor Temperature Field

As an important content of motor design, the theory and research for motor temperature field calculation are constantly improving with the upgraded motor manufacturing. Since the 1970s, modern numerical methods and applications of computers lay a solid foundation for the study of temperature field and make it flourish with a relatively comprehensive theoretical system and calculation methods. Some commonly used methods so far are mainly shown as the following:

4.1.2.1 Simplified Formula Method

The simplified formula method is relatively simple, and it is a temperature-rise estimation method commonly used in the design of motor manufacturers to calculate the average temperature rise of copper and iron-cores of armature windings. The simplified formula assumes that all core loss and the copper consumption of the effective part is dispersed only through the cylindrical cooling surfaces of stator or rotor, while there is no heat transfer between the effective and terminated parts of the armature winding copper. Although these assumptions are not fairly reasonable, the thermal coefficient used in this method is based on the results of the motor temperature rise experiment with the same or similar structure, so the calculation results are rather close to the actual.

4.1.2.2 Equivalent Thermal Circuit Method and Equivalent Thermal Network Method

The basic principles of the equivalent thermal circuit method and the equivalent thermal network method are the same. They are both equivalent heat circuit which are based on heat transfer and circuit theory. They adopt local concentration parameters, assuming that the heat source loss is concentrated in each node, which are joint by heat resistance. The concentrated heat flux flow through the branches, taking the node temperature as the dependent variable. The equivalent thermal network is composed of the loss, heat flux, heat resistance, heat capacity, and the temperature rise of certain points. According to Kirchhoff's law of heat flux, the equations of temperature of the network nodes are determined and the mathematical-physical model is established. With a numerical solution, the algebraic equations regarding the temperature values of each node can be calculated. Because the thermal circuit follows similar physical laws to the circuit, more mature circuit technology is applicable.

The difference between the two methods is only the degree of simplification of the heat transfer of the actual motor. The thermal circuit method includes fewer nodes and the calculation, while the thermal network method involves more nodes, which provide a more comprehensive consideration to the factors affecting motor heating, heat transfer and heat dissipation, therefore the calculation results are more accurate. As a result, foreign literature often does not make a special distinction between the two methods. They are collectively named Lumped Parameter Thermal Network (LPTN).

4.1.2.3 Numerical Methods

The heat transfer in the motor can be expressed by the thermal conduction differential equation according to the ordinary law of heat exchange. Numerical method is to solve the differential equation by modern numerical method by dividing the solution area into many small units and establishing the equation in each unit, therefore the whole equations have been solved. Finite Element Method (FEM) and Computational Fluid Dynamics (CFD) are the most popular ones in the relative literature, details are shown as the following.

(1) Finite Element Method (FEM)

The finite element method is a numerical method of solving mathematical physics problems according to the principle of variable fraction. In the motor, the heat flow that moves from one effective part to another is actually non-existent, so the finite element method is solving the pending variables at the finite node through an idealized collection of limited units of the motor that are not related to each other in terms of heat and in a relatively simple geometric shape.

The biggest advantage of finite element method is that the temperature field model of any structure motor can be established and solved by geometric modeling and mesh generation, the intuitive cloud map of temperature distribution can be obtained, and the value and position of the overheating point are given. The disadvantage is that the estimation of convective heat transfer coefficients as boundary conditions still need to be carried out through empirical formula, therefore the accuracy is lowered.

With the enhancement of computer computing speed and 3D mapping ability, computing and modeling are no longer problems, and boundary conditions can be calculated accurately in combination with other methods. In this way, finite element method is widely used as a temperature field calculation method.

(2) Computational Fluid Dynamics (CFD)

Computational fluid dynamics is a new discrete method based on classical fluid dynamics and numerical calculation methods. Through computational and numerical calculation and image display, the numerical solution of the flow field in time and space can be described quantitatively, so as to study on the physical problems. It provides effective computing technology for many complex flow and heat transfer problems in modern science.

With CFD method, not only can the energy conservation equation be established in the solids and fluids in the solution domain, the mass conservation equation and momentum conservation equation of the fluid can also be established separately, in order to describe the speed field and pressure field of the cooling medium, and the convection heat coefficient of the motor can be calculated by the heat differential equation, so as to calculate the temperature field of the motor accurately. Its shortcoming is that it requires too much calculation, the strong hydrodynamic basis is needed, and it's more difficult to converge the calculation results.

Firstly, compared with conventional motors, the pulsed alternator has a special structure, in which the rotor's outer edge is equipped with a compensation aluminum shield. To overcome the centrifugal force and radial elongation of the winding during high-speed rotation, carbon fiber is winded outside the rotor slotless excitation winding; Secondly, it works in a special condition called pulsed discharging. In a few milliseconds, thousands or even hundreds of thousands of pulsed current can be produced. The heat is generated instantly when discharging and dissipated during the interval. Therefore, in order to accurately calculate the influence of the frequency and times of the pulses on the motor temperature, it is necessary to simulate its transient temperature field.

Combining the advantages and disadvantages of several temperature field calculation methods, and taking into account the unique air-core structure and pulse condition of the air-core pulsed alternator, this book adopts the numerical finite element method for its advantages of establishing the simulation model accurately according to the engineering drawings, establishes the three-dimensional temperature field model of the pulsed alternator, and brings about a simulated computation of the transient temperature field.

4.1.3 Air-Core CPA Temperature Field Analysis

4.1.3.1 Basic Assumptions and Solution Domain Models

Taking a stator double armature air-core compensated pulsed alternator as an example, the temperature field analysis is carried out as following. The three-dimensional section view and radial section of the motor are shown in Figs. 4.1 and 4.2 respectively.

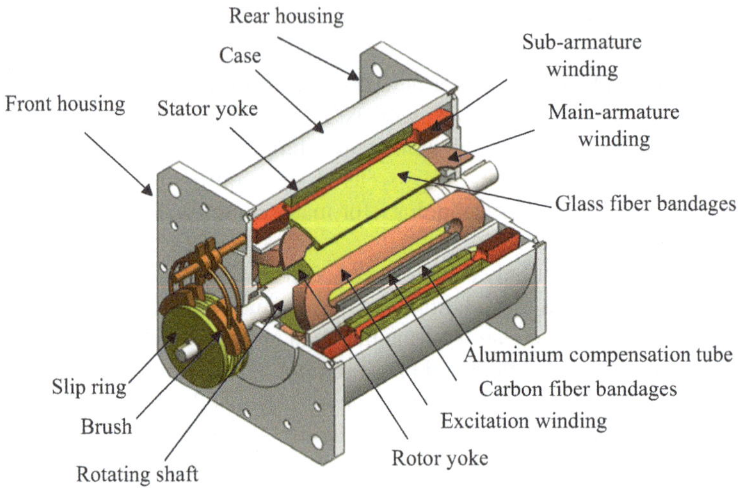

Fig. 4.1 DSW-PCPA 3D Section view

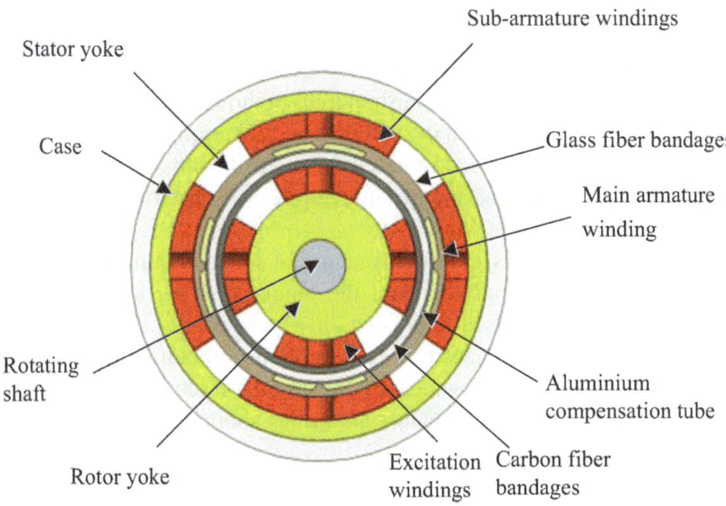

Fig. 4.2 DSW-PCPA radial cross-section

The principle prototype of air-core compensated pulsed alternator prototype is a single-phase, four-pole, passive compensation structure, the stator uses two sets of windings namely the main armature and sub-armature. According to the respective working characteristics of the two sets of armature windings, the two windings are designed to optimize the performance of the motor. The iron-core machine adopts slotless winding to reduce the slot and addendum leakage inductance of the winding, thereby reducing the total inductance. In the air-core machine, the stator yoke is made

Table 4.1 Parameters for Double Armature Air-core Pulsed Alternator

Mechanical parameters		Electromagnetic parameters	
Rated speed (rpm)	10 000	Number of phases	1
Moment of inertia(kg·m^2)	0.086	Number of poles	4
Rotor Inertial Energy Storage (kJ)	47.2	Main armature winding no-load voltage peak (V)	450
Motor quality (kg)	60	Sub-armature winding no-load voltage peak (V)	4 400
Motor length (mm)	350	Discharge current peak (kA)	3.5
Motor Outer Diameter (mm)	290	Excitation current peak (A)	200
Rotor Outer Diameter (mm)	172	Number of turns per pole on main armature windings	10
Aluminium shield length (mm)	312	Number of turns per pole on sub-armature windings	250
Aluminium shield thickness (mm)	6	Number of turns per pole on excitation armature windings	180

of non-magnetic fiber resin composite material, and the inductance of the winding is not affected by the slotting. However, if the structure made of fiber resin composite materials is slotted, the fiber will be cut off, so that the strength of the structural parts will be greatly reduced. Therefore, in order to achieve the overall molding of the stator and rotor components, and enhance their overall strength, the slotless windings are still adopted in the air-core motor. The prototype parameters are shown in Table 4.1.

According to the actual structure of the motor and the theory of heat transfer, the following assumptions and simplifications are made to the temperature field model of the pulsed alternator:

1. The mechanical loss is ignored;
2. The resistivity and the thermal conductivity of the material are ignored with the change of temperature;
3. The air temperature in the motor is evenly distributed, and the air temperatures of the stator, rotor, air gap and end are the same;
4. The contact with thermal resistance is ignored, as the components in the air-core CPA are bonded by epoxy resin and there is no assembly gap, so the resin can be modeled together as a composite material with similar thermal conductivity;
5. The motor structure has periodic symmetry along the circumference direction, and is completely symmetrical along the axis on both sides of the central cross-section, so it is desirable to use the half-pole under-motor model as the solution domain, simplifying the simulation model to 1/16 of the actual model for calculation.

Based on the above hypotheses, the model of the temperature field solution of pulsed alternator is established as shown in Fig. 4.3.

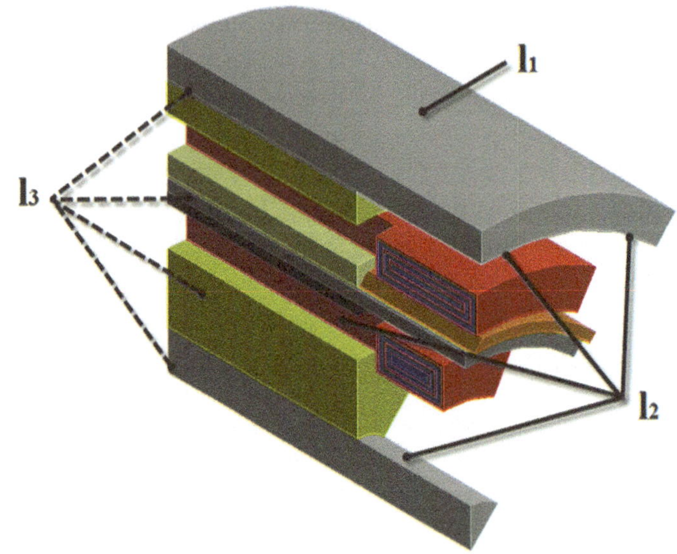

Fig. 4.3 Solving domain models and boundary conditions

In Fig. 4.3, area l_1 is the surface of the case exposed to external air, dissipating through natural convection and radiation; area l_2 is the surface where the motor is in contact with the internal air, such as the end of the winding, the end of the stator yoke and rotor yoke, the gap between the poles of concentric winding etc. These areas carry out heat transfer with the internal air by natural convection or forced convection which is also an important component of motor heat dissipation; area l_3 is symmetrical, which is set as an insulation boundary condition.

4.1.3.2 Winding Model

It is almost impossible to model the wires one by one when building a winding thermal analysis model, therefore the winding model must be under equivalent simplification. Common equivalent winding models can be classified into two main categories:

The first one classifies windings, insulation and air as a thermal conductor according to the actual rate of fullness of winding slot and impregnation conditions, It also calculates the average thermal conductivity of the equivalent thermal conductor according to Fourier's law.

Such winding model is simple and can express a certain temperature gradient. The windings are thermally conductive on multi-layer plane-wall radially and circumferentially, and the equivalent thermal conductivity can be solved by the following formula:

$$\lambda_{eqr} = \frac{\sum_{i=1}^{n} \delta_i}{\sum_{i}^{n} \frac{\delta_i}{\lambda_i}} \tag{4.4}$$

λ_{eqr} the radially equivalent thermal conductivity of the winding in W/(m·K);
δ_i the equivalent thickness of the layer i of the material in m;
λ_i average thermal conductivity of layer i of the material, in W/(m·K).

The axially thermal conductivity of the parallel connection and combination wall of the conducts can be solved as follows:

$$\lambda_{eqa} = \frac{\sum_{i=1}^{n} \lambda_i S_i}{\sum_{i=1}^{n} S_i} \tag{4.5}$$

λ_{eqa} the axial equivalent thermal conductivity of the winding in W/(m·K);
S_i equivalent area of layer i of the material in m^2;
λ_i average thermal conductivity of layer i of the material W/(m·K).

As can be seen from the above two equations, the winding equivalent thermal conductivity calculated by this algorithm is known as anisotropy, and the axially equivalent thermal conductivity is often hundreds of times as the radially and circumferentially equivalent thermal conductivity. In the three-dimensional finite element model, it's difficult to express for the anisotropy material at the end corner, so this method is only suitable for two-dimensional simulation model, which is not suitable in this model.

The second one is to treat all the copper wire of the winding as a thermal conductor, and all insulation materials of the winding are deemed to be equivalent to another thermal conductor, which is modeled separately, so that there is no three-dimensional modeling problem brought about by the anisotropy of the thermal conductivity. However, the thermal conductivity of copper is very large, while the thermal conductivity of insulation is very small. Based on the temperature field calculation results, the temperature of the entire equivalent copper conductor is often the same, while the temperature of the equivalent insulation material varies greatly. The temperature gradient is displayed on the smaller geometry of the insulator. To accurately calculate winding temperature, it's necessary to refine the mesh of the insulator, but it will also increase the complexity of the simulation model. at the same time.

Therefore, according to the advantages and disadvantages of the above two methods, the equivalent winding simulation model is established as shown in Fig. 4.4, according to the volume of the actual winding and the proportional relationship between copper and epoxy resin and insulating paint in the winding. The winding model is composed of three layers of copper conductor and three insulators, which reflects a certain temperature gradient in the winding without making the computation of the model more challenging.

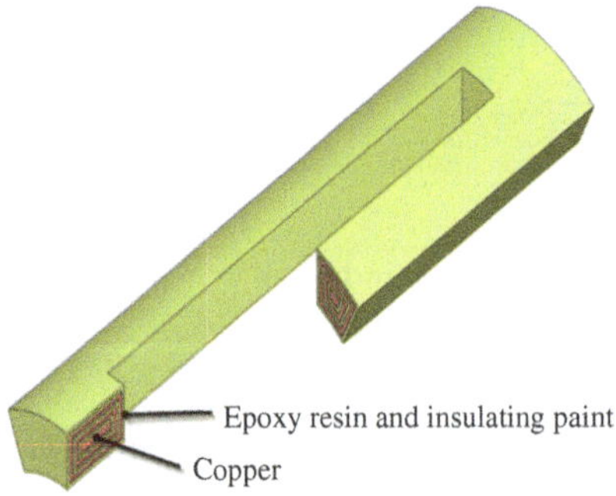

(a) Equivalent winding model

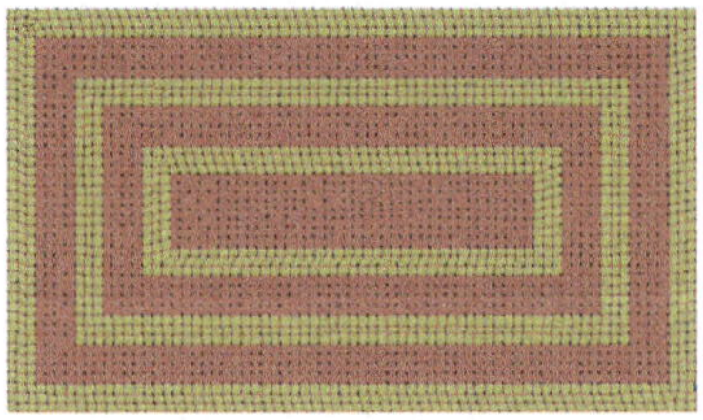

(b) Winding cross-section

Fig. 4.4 Equivalent winding model and section

4.1.3.3 Air Gap Equivalent Model

As a medium for heat exchange between stators and rotors of the motor, the air gap often requires exclusive treatment in thermal analysis models. The rotation of the rotor drives the air flow in the air gap, so that the heat exchange between the stator and the air gap and between the rotor and the air gap is mainly convective. The air in the air gap, as a heat transfer body, has conductive heat exchange between the rotor and stator, and the air gap temperature increases in during the operation of the motor until it becomes steady.

Finite element software generally cannot handle the internal boundary condition, that is, in the finite element model of the motor, if the air gap model is established to connect with the stator and rotor, the convective heat exchange boundary condition cannot be applied to the inner edge of the stator and the outer edge of the rotor. Therefore, if the motor temperature field is calculated by finite element method, the common method is to separate the stator and the rotor, apply convection boundary conditions on the inner edge of the stator and the outer edge of the rotor, and then conduct thermal analysis separately. However, another problem comes with this method is that air temperature in the air gap is difficult to determine and needs to be determined in combination with computational fluid mechanics (CFD) or experimental measurements. CFD increases the complexity of the model and requires a solid basis of fluid mechanics, while the experimental measurement cannot be applied in the design of the motor because it's not qualified enough to be a guide for it.

In this view, the book uses the effective thermal conductivity coefficient of air gap for the research. The thermal coefficient of the static fluid is used to describe the heat exchange ability of the flowing air in the air gap, i.e. the heat transmitted by the stationary fluid is equal to the heat transmitted by the flowing fluid, so that the rotating rotor can be treated as stationary. After the equivalent treatment, it can be regarded that the heat exchange between stator and rotor is completely carried out by air gap conduction. Convection boundary conditions, are no longer needed, therefore solves the problem that the software cannot deal with the internal boundary conditions.

The effective coefficient of thermal conductivity can be extracted from the formula below,

$$\lambda_{eff} = 0.069 \cdot \eta^{-2.9084} \cdot Re^{0.4614 \cdot \ln(3.3361 \cdot \eta)} \tag{4.6}$$

λ_{eff} the effective thermal conductivity of the air gap in W/(m·K);

$$\eta = \frac{r_0}{R_i} \tag{4.7}$$

r_o outer radius of rotor in m;
R_i radius within stator in m;

$$R_e = \frac{r_0 \omega_m \delta}{\upsilon} \tag{4.8}$$

R_e Reynolds Number;
ω_m rotor rotation angle speed in rad/s;
δ air gap length in m;
υ the motion viscosity of the air in m^2/s.

4.1.3.4 Determination of the Heat Transfer Coefficient of the Surface

The heat transfer coefficient of the surface is the coefficient of heat emission, and its size is related to many factors in the heat transfer process. It depends on the properties of the fluid and the shape, size and arrangement of the heat-exchange surface. It also relies greatly on velocity. The heat transfer coefficient of the surface can be expressed as,

$$h = f(u, \lambda, c_p, \rho, \eta, \gamma, T_f, T_w, L, \phi \ldots) \tag{4.9}$$

u the velocity of the fluid;
λ the thermal conductivity of the fluid;
c_p heat capacity at constant pressure;
ρ density;
η the dynamic viscosity of the fluid;
γ volume expansion coefficient;
T_f the temperature of the fluid;
T_w the temperature of the heat dissipation wall;
L and ϕ the size and shape of the system.

Table 4.2 Heat transfer coefficient on the surface of the case when the air is static

Features of the surface	h (W/(m^2·K))
Painted-only cast iron or steel surfaces	16.7
Puttied and painted iron or steel surfaces	14.2
Painted surface of copper	13.3

As can be seen from Eq. 4.9, heat transfer coefficient of the surface is a complex function relying on multivariate variables, which is not quite suitable for accurate theoretical calculation, and is a challenge in the study of temperature field. In engineering, the empirical formula method is generally used, and under the guidance of similarity theory, the functional relationship between similarity criteria is established through experiments. This section refers to the surface heat transfer coefficient of induction motors and synchronous generators to calculate the empirical formula and does some correction to calculate the surface heat transfer coefficient of each component of the air-core pulsed alternator.

(1) Heat transfer coefficient on the surface of the casing

The cooling of air-core CPA model prototype is realized by natural convection. During discharging, heat of the windings and compensation aluminum shield exchanges with the surrounding environment mainly through the thermal convection and thermal radiation on the surface of the casing. According to the relevant motor manufacturing experience, the natural convective heat exchange coefficient can be multiplied by the correction coefficient to characterize the combined effect of convection and radiation. Commonly used motor surface characteristics and the corresponding surface heat transfer coefficients are as shown in Table 4.2, wherein, the prototype case is paint-only steel surface, so the corresponding heat transfer coefficient is 16.7 W/(m^2·K).

(2) Heat transfer coefficient of end surface

As mentioned earlier, in order to compensate for the armature reaction at the end of the winding, the compensation aluminum shield on the model prototype rotor extends all the way to the outer edge of the end of the main armature winding. When the motor rotates, the end is also in a forced convection heat exchange state under the influence of the rotation of aluminum shield. Since there is no axial ventilation in the motor, it is possible to calculate by reference to the thermal coefficient at the end of the fully enclosed induction motor.

According to the fluid similarity theory, for forced convection cooling, the thermal coefficient h can be obtained by the following equations:

$$\begin{cases} h = \frac{N_u \lambda_a}{L} \\ N_u = a(R_e)^b (P_r)^c \\ R_e = \frac{vL}{\upsilon} \end{cases} \tag{4.10}$$

N_u Nusselt number;
R_e Reynolds number;
P_r Prandtl number;
v characteristic velocity in m/s;
L characteristic length in m;
λ_a the thermal conductivity coefficient of air in W/m·K;
υ the kinematic viscosity of the air in m^2/s.

Considering that the rotor end ring of the model prototype does not contain fan leaves, the concept of fan efficiency is introduced, the characteristic velocity in the Reynolds equation can be obtained according to the following formula,

$$v = \frac{2\pi n r_o}{60}\eta \tag{4.11}$$

n rotating speed in rpm;
η fan efficiency, which can be modified experimentally

(a) Heat transfer coefficient at the surface of the end of the winding

The model prototype belongs to the closed motor, in which the internal and external air flow do not circulate. This gives rise to poor cooling conditions of the end winding. The inner and outer surfaces of the winding are described by different criteria equations due to their natural differences.

The inner surface of the winding end,

$$N_u = 0.103 R_e^{\frac{2}{3}} \tag{4.12}$$

The outer surface of the end of the winding,

$$N_u = 0.456 R_e^{0.6} \tag{4.13}$$

In the formula, the characteristic length L is the equivalent diameter of the corresponding winding end, and the heat transfer coefficient of the surface of each winding end can be obtained under a given rotational speed by inputting in Formula 4.10.

(b) Heat transfer coefficient on the surface of the yoke surface

The end surface of the stator yoke mainly carries out convective heat loss by formed by the flow of air in the machine, the thermal coefficient is not only related to the surface air velocity, but also related to the state of the flow. According to the heat exchange experimental formula of the Y2 series stator induction motor, the convection coefficient of the stator yoke and the air in the machine can be calculated,

$$h = 15 + 6.5 v^{0.7} \tag{4.14}$$

v effective linear velocity in m/s.

There is also convective heat exchange between the surface of the end of the rotor yoke and the in-machine air. It can be expressed in the criterion equation as

$$N_u = 1.67 R_e^{0.385} \tag{4.15}$$

L outside radius of the rotor.

(3) Surface heat transfer coefficient between windings

As mentioned earlier, the excitation windings and sub-armature windings of the model prototype are constructed with concentric winding structures, and there is a certain gap between the wire packs of the winding. In the gap, the convective heat exchange exists between the air in windings and motors, similar to the axial ventilating duct of the motor.

For the gap between the wire packs of the sub-armature winding of the stator, since the prototype is completely enclosed and does not contain any additional cooling equipment, it can be considered that it belongs to the natural convection heat exchange in limited space, the Nusselt criteria can be obtained according to the following formulae.

$$G_r = \frac{gL^3\gamma\Delta T}{\upsilon^2} \tag{4.16}$$

$$\begin{cases} N_u = 0.59(G_r P_r)^{0.25} & 1.43 \times 10^4 \le G_r P_r \le 3 \times 10^9 \\ N_u = 0.0292(G_r P_r)^{0.39} & 3 \times 10^9 \le G_r P_r \le 2 \times 10^{10} \\ N_u = 0.11(G_r P_r)^{0.33} & G_r P_r > 3 \times 10^9 \end{cases} \tag{4.17}$$

G_r Grashof number;
g acceleration of gravity in m/s^2;
L characteristic length in m;
γ coefficient of cubic expansion, $\gamma = 1/(273 + T_f)$, in 1/°C;
T_f the temperature of the fluid;
ΔT the difference between fluid and solid, in °C.

For the gap between the rotor excitation winding line pack, the rotor's rotation has a significant influence on the fluid resistance and cooling in these "ventilating ducts". The effect is related to how do centrifugal force and Coriolis force impact the flow of the cooling medium. Therefore, it belongs to the heat exchange problem in the ducts where the cooling medium rotates around the outer center line instead of doing forced flow.

Studies have shown that even if air does not move by force through the duct, the axial ventilating duct in the rotor of the motor still imposes a significant effect on the temperature field. Closing these ducts will increase the rotor temperature by about

10–12 °C, and the stator temperature by 6 ~ 8 °C. This is due to the air convection occurred in the axial duct as the rotor rotates.

The heat of the winding is passed through the "wall of the ventilating duct" to the air, which increases the temperature and decreases the density. Under the action of centrifugal force, hot air is pressed by cold air to the lower part of the shield, and then squeezed out by a large amount of new preheating gases, outflows through the opening end of the "ventilating duct" therefore the convection is formed. In this case, the heated light gas flows in the opposite direction to the centrifugal force. This effect is the same as the natural convection enhanced by the force of gravity field in a fixed preheating short ventilating duct. However, centrifugal acceleration is much larger than gravity acceleration, so the natural convection in the rotating ventilating duct is much more intense than the fixed one.

In order to calculate the surface heat transfer coefficient in the rotating duct, a similar criterion equation obtained when studying the cooling of air under natural convection conditions can be applied,

$$\begin{cases} G_r^* = \frac{fL^3\gamma\Delta T}{\upsilon^2} \\ N_u = 0.47G_r^{*0.25} \end{cases} \tag{4.18}$$

G_r* the modified Grashof number

$$f = \omega_m^2 r_o \tag{4.19}$$

f centrifugal acceleration, in m/s$^{2;}$
ω_m rotor rate of angular motion in rad/s;
r_0 outer radius of the rotor in m

Using the calculation formula of the heat transfer coefficient of the surface listed in this section, the coefficient of each component is calculated at the design speed of 10 000 rpm, as shown in Table 4.3.

Table 4.3 Surface heat transfer coefficient at 10 000 rpm (W/(m^2·K))

Components	Sub-armature windings Outer surface of the end	Sub-armature windings Inner surface of the end	Main armature winding Outer surface of the end	Main armature winding Inner surface of the end	Excitation windings Outer surface of the end
h	164.47	89.14	174.06	93.45	200.01
Components	Excitation windings Inner surface of the end	Surface of the stator yoke	Sub-armature windings Surface of the wire pack	Surface of the end of the rotor yoke	Excitation windings Surface of the wire pack
h	104.92	108.4	11.82	62.36	59.62

4.1.4 Example of Air-Core CPA Temperature Field Analysis

This section takes the 1.1-meter orbital gun as the discharge load to simulate the air-core CPA load temperature field. In simulation, assuming that the orbital gun is the ideal load and the parameters of the continuous emission of projectile remain unchanged, the ability of the system to achieve continuous discharge operation only depends on whether the air-core CPA exceeds the temperature limit.

4.1.4.1 Air Temperature Determination in the CPA

The air temperature in the motor is a vital parameter in the calculation of the motor temperature field, which is not only related to the material parameters of the fluid, but also as the third type of boundary condition of heat exchange, which directly affects the temperature distribution in the motor. However, it is difficult to give the exact value of the air temperature in the motor by theoretical calculation. The calculation of conventional motor air temperature is mainly based on engineering experience. For special motors, it is essential to calculate by iteration. Because the model prototype lacks sufficient experimental basis, this book uses the iterative method, aiming at the average temperature of the equivalent air gap, to solve the air temperature in the motor, and the solution process is shown in Fig. 4.5.

1. Given the initial air temperature value, and the material parameters of the air are determined by this value, the surface heat transfer coefficient of each component is obtained;

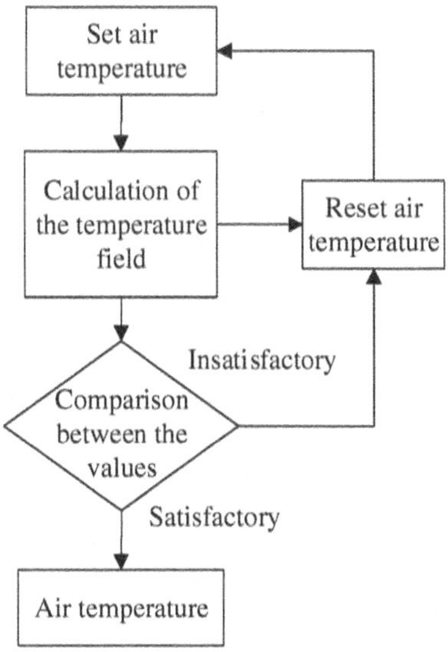

Fig. 4.5 Calculation flowchart of the air temperature in the motor

2. To obtain the temperature field of the prototype with the calculated surface heat transfer coefficient, taking air temperature as the boundary condition;
3. As mentioned earlier, as a result of the equivalent treatment of air gaps, air gap air is no longer a convection boundary condition, but rather a thermal conductor as part of the model, thus seeking its average temperature;
4. Compare the difference between the given air temperature and the average temperature of the equivalent air gap. If it is more than 10%, the air temperature needs to be re-set; If it is less than 10%, the iterative process can be terminated, and the result should be taken as the temperature in the motor.

4.1.4.2 Heat Source Computation

Because the model prototype uses a full-air-core structure, the entire motor does not contain ferromagnetic material, there is no iron consumption in the motor. When discharging, the heat source mainly comes from the copper consumption generated by the discharge current of the main armature winding, which compensates for the eddy current loss of that the aluminum shield inducted, and the copper consumption from the excitation winding and self-exciting process of the sub-armature winding.

(1) Calculation of heat generation rate of the main armature windings and aluminum shield

Based on the electromagnetic field finite element method, the average heat generation rate of the main armature winding is 4.5×10^8 W/m^3 within 2 ms. However, if taking the thermal rate as a pulsed heat source, it will be difficult for temperature field simulation. When there are more discharges, the pulsed heat resource needs to be reloaded on simulation model, therefore simulation time step will also be greatly limited, and simulation efficiency is not desirable. Therefore, with a simplified method, this book uses the average heat generation rate in a discharge cycle instead of pulsed heat generation rate as a heat source applied to the simulation model of windings. In the case of a 30-second long discharge cycle, the average heat generation rate h_{pa} in one cycle duration in the main armature winding model should be,

$$h_{pa} = \frac{4.5 \times 10^8 \times 2 \times 10^{-3}}{30} = 3 \times 10^4 \mathrm{W/m^3} \tag{4.20}$$

Similarly, it can be calculated when the discharge cycle is 30 s, the average heat generation rate of the aluminum shield is 9.7×10^3 W/m^3 in one cycle duration.

(2) Calculation of heat generation rate of excitation windings and sub-armature windings

In the model prototype self-excitation is adopted, and the circuit is shown in Fig. 4.6. At the beginning of the self-excitation, the capacitor provides a seed current to the excitation windings, and the sub-armature winding inducts a sine voltage under the action of the main magnetic field. The excitation winding is induced back through a

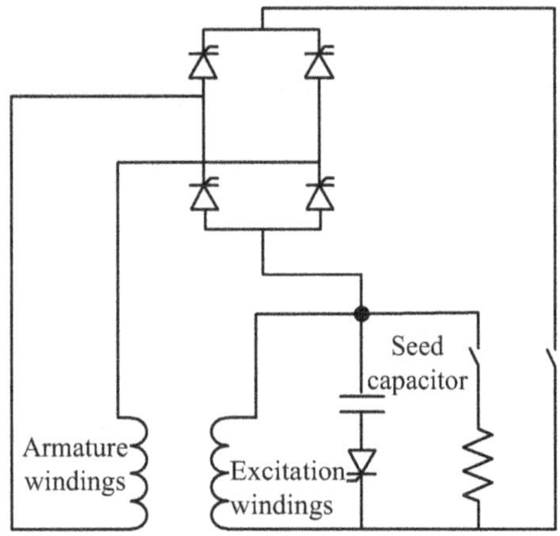

Fig. 4.6 Schematic diagram of self-excitation circuit

single-phase full-wave rectifier circuit with capacitive filtering, resulting in a larger DC excitation current, forming positive feedback, and increasing the excitation current. Finally, the rated excitation current reaches 200 A.

This section mainly studies on the temperature field of the air-core CPA, therefore simplified the complex self-stimulation process. Ignoring the rise of excitation current during self-excitation, assuming that excitation time is set as 0.1 s in each loaded discharging process, during which the excitation current remains to be 200 A, then the average heat generation rate h_f in a discharge cycle of the excitation winding is,

$$h_f = \frac{\rho J^2 t_f}{T} \tag{4.21}$$

ρ copper resistivity in Ω·m;
J current density, in A/m^2;
t_f excitation time, in s, (0.1 s here);
T discharge cycle, in s.

After making the simplified hypothesis of the self-excitation process, according to the principle of single-phase full-wave rectifier circuit with capacitive filtering, the effective value of the sub-armature winding current is 221A in the 0.1-second long excitation time. Similarly, the average heat generation rate of the sub-armature winding in a circle duration.

According to the above analysis, giving 15 s, 30 s, 60 s, the three typical discharge cycles of CPA, the average heat generation rate of each winding and aluminum shield can be calculated and summarized as shown in Table 4.4.

Table 4.4 Average heat generation rate in a discharge cycle (W/m^3)

Serial number	Cycle (s)	Main Armature	Sub-Armature	Excitation	Aluminum shield
1	15	6.0×10^4	1.80×10^6	1.74×10^6	1.93×10^4
2	30	3.0×10^4	8.99×10^5	7.36×10^5	0.97×10^4
3	60	1.5×10^4	4.49×10^5	3.68×10^5	4.83×10^4

4.1.4.3 Calculation of Load Temperature Field

This section is based on the model prototype temperature field simulation model, as shown in Fig. 4.3. The analysis and calculation of three different discharge cycles (15 s, 30 s, 60 s) of the temperature field of the air-core CPA are carried out. The simulation of the model prototype is designed at the rotational speed of 10,000 rpm, at which point the thermal coefficients of each component are shown in Table 4.1. Since the epoxy resin in the model prototype will accelerate aging when the temperature exceeds 100 degrees C, this should be used as a temperature limit for the number of discharges of the CPA. The initial temperature of the simulation is assumed to be 25 degrees at room temperature.

When the length of load discharge cycle is 15 s, as shown in Fig. 4.7, it is the 285 s of the transient simulation, that is, after discharging 19 times, the temperature of sub-armature winding at the motor hotspot exceeds 100 °C, at this time the motor must be shut down for a while for cooling in order to continue the load

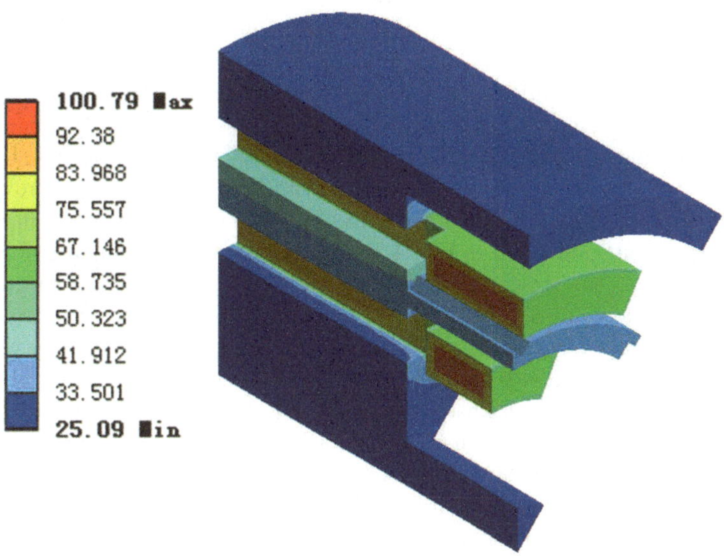

Fig. 4.7 After 19 discharges (285 s) the temperature distribution cloud picture of the air-core CPA when the length of the circle is 15 s

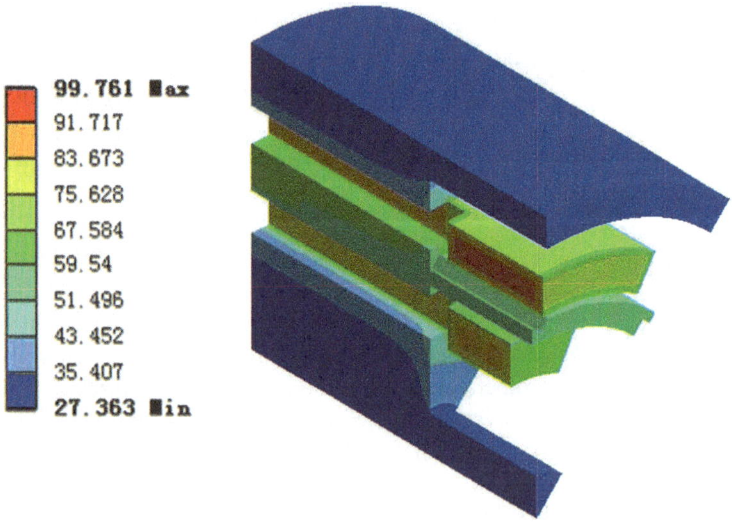

Fig. 4.8 After 30 discharges (900 s) the temperature distribution cloud picture of the air-core CPA when the length of the circle is 30 s

discharge. From the temperature distribution cloud picture, it can be obviously seen that the temperature distribution in the motor is very uneven. The temperatures of the sub-armature winding and excitation winding are significantly higher than other components, this is because the composite material is a poor conductor of heat, and the thermal conductivity is very low. It is difficult for the heat source winding steam to get effective heat dissipation. With the increasing number of discharges the heat continues to accumulate inside the winding, leading to a rise in the winding temperature.

When the length of load discharge cycle is 30 s, as shown in Fig. 4.8, it is the 900 s of the transient simulation, that is, after discharging 30 times, the temperature of sub-armature winding at the motor hotspot is 98 °C, which is about to exceed the upper temperature limit. Compared to Fig. 4.5, the temperature distribution at this time is more uniform. The temperatures of rotor carbon fiber bandages, compensation aluminum shield, equivalent air gap, the main armature winding and stator fiber glass fiber bandage have increased significantly, with an also remarkably reduced difference between the temperature of excitation winding and sub-armature winding.

When the length of load discharge cycle is 60 s, as shown in Fig. 4.9, it is the 3120 s of the transient simulation, which means after discharging 52 times, the temperature of sub-armature winding at the motor hotspot is 99.8 °C. Compared with the first two operating conditions, the temperature distribution in the motor is more uniform at this time, even the lowest temperature in the motor reached 46.2 °C. The heat dissipation of the heat source armature is more adequate, and the number of discharges has been significantly increased.

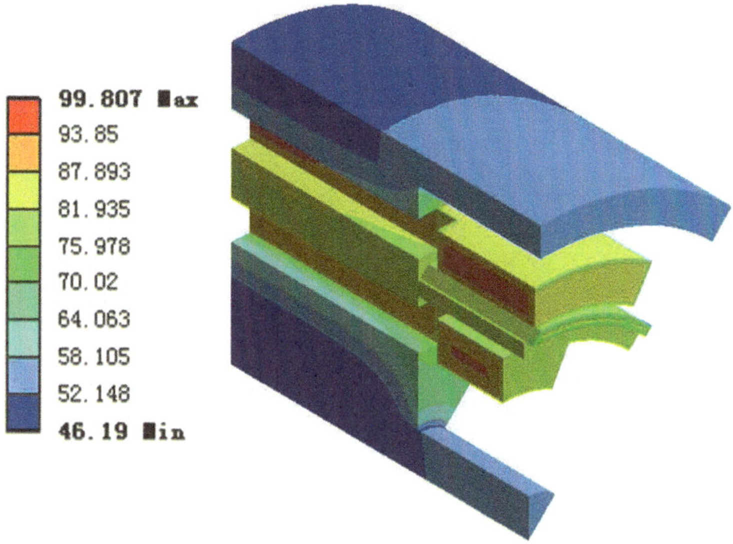

Fig. 4.9 After 52 discharges (3120 s) the temperature distribution cloud picture of the air-core CPA when the length of the circle is 60 s

The load operating time is converted into discharge times in a curve as shown in Fig. 4.10, which can express the relationship between the number of discharges and the temperature of the hottest spot of the windings under different discharge cycles. It can be analyzed that as the discharge cycle increases, the interval between the discharging pulse and the pulse increases, the winding heat dissipation is longer, so the number of pulses required to achieve the temperature limit is also increased. This implies that the number of load discharges allowed by the air-core CPA increases with the extension of discharging cycle.

It can also be seen from the figure that with the increase of discharging cycle, the slope of the temperature rise of the hottest spot of the winding decreases significantly. When the cycle is 60 s, the winding temperature becomes closer to the steady state with the growing number of discharges. It can be inferred that when the discharge cycle continues to increase to a certain value, the winding temperature can reach a steady state at the temperature limit of 100 °C. It can be deducted that at this temperature, the heat generated by the winding in the moment of discharge can be completely dispersed at the discharge interval.

According to the above analysis, the discharge cycle needs to be extended, and a steady state analysis needs to be carried out for a loaded air-core CPA temperature field simulation model, until the temperature of the hottest spot of the steady state winding below the temperature limit of 100 °C. The calculation shows that when the discharge cycle is raised to 120 s, as shown in Fig. 4.11, as the motor temperature

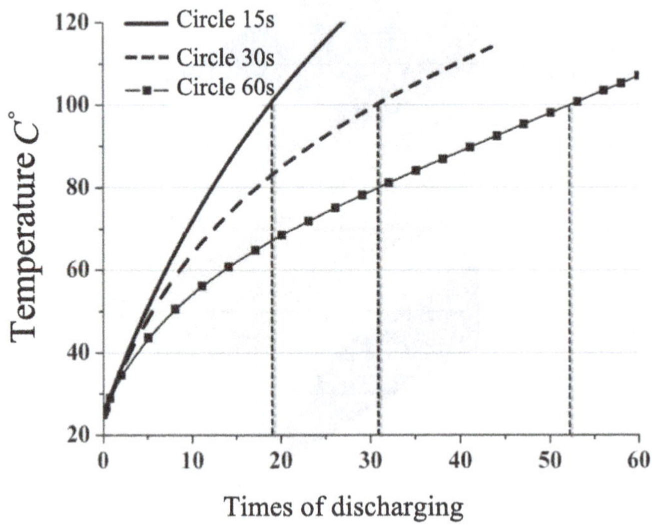

Fig. 4.10 The number of discharges and the temperature of the hottest spot of the windings in different discharge cycles

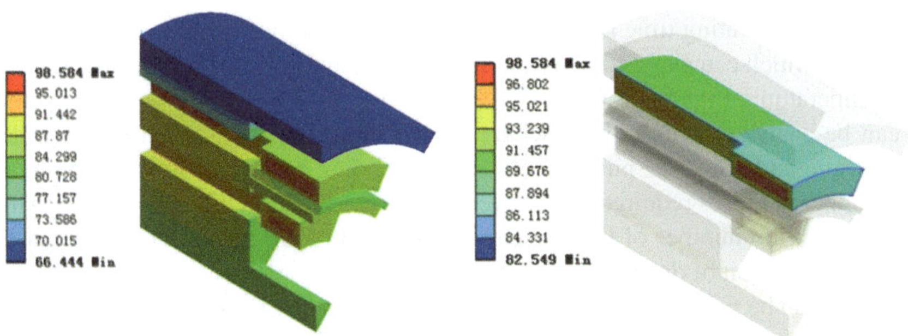

(a) Air-core CPA domain (b) Sub-armature winding

Fig. 4.11 Cycle 120 s, Steady-state air-core CPA temperature distribution cloud

reaches a steady state, the temperature of the hottest spot of the winding is 98.6 °C, which is below the temperature limit of epoxy resin. This exemplifies that as an ideal 1.1-meter long orbital gun load of pulsed power supply, if the continuous emissions of projectiles are required such prototype, discharge cycle must be longer than 120 s.

4.2 The Computation Basis of the Pulsed Alternator Cooling

4.2.1 The Basis of Computational Fluid Dynamics

Conjugate heat transfer occurs in the cooling system, that is, water cools the heat source winding but also being heated at the same time. There is a rise in the temperature of the water from the inlet to the outlet along gradually. The temperatures of the winding and cooled water affect each other, so it is difficult to solve this problem from the perspective of heat transfer alone. Based on computational fluid dynamics (CFD), this section uses the simulation software Ansys CFX to establish a simulation model of the cooling structure, designs the cooling structure from the view of fluid and heat transfer conjugate, and briefly introduces the basis of computational fluid dynamics and basic fluid control equations.

4.2.1.1 Overview of Computational Fluid Dynamics

Computational fluid dynamics is an analysis of systems containing related physical phenomena such as fluid flow and thermal conduction through numerical calculations and image display. The idea of CFD boils down to replace the continuous physical quantity field in the time and space domain originally, such as the velocity field and the pressure field, with a set of variables on a limited number of discrete points; and to establish the algebraic equations about the relationship between the field variables on these discrete points by certain principles and methods, then solving the algebraic equation system to obtain the approximate value of the field variable.

CFD can be regarded as a numerical simulation of flow under the control of the basic equation of fluid. Using this numerical simulation, we can obtain the distribution of the basic physical quantities (e.g. velocity, pressure, temperature, concentration, etc.) at various locations in the flow field of extremely complex problems, and how these physical quantities change over time.

4.2.1.2 Fluid Dynamics Governing Equations

Flow of the fluid is regulated by the law of physical conservation, and the basic law of conservation includes the law of mass conservation, the law of constant momentum, and the law of energy conservation. Governing equations are mathematical descriptions of these laws of conservation.

1. Mass conservation equation

Any flow must meet the law of quality conservation. The law can be expressed as: the increase in mass in the fluid micro-unit in a unit of time is equal to the net mass flowing into the micro-unit within the same time. Depending upon this law, a mass conservation equation can be derived:

$$\frac{\partial \rho}{\partial t} + \frac{\partial (\rho u)}{\partial x} + \frac{\partial (\rho v)}{\partial y} + \frac{\partial (\rho w)}{\partial z} = 0 \tag{4.22}$$

ρ mass density;
t time;
u, v and w on the field variables.

2. Momentum conservation equation

The law of constant momentum is also the basic law that is applicable to any fluid system. The law can be expressed as follows: the rate of change of the dynamic of the fluid in the micro-meta body is equal to the sum of the various forces on which the external body acts. This law is actually Newton's second law. For Newtonian fluids, the momentum conservation equation is:

$$\frac{\partial (\rho u)}{\partial t} + \mathrm{div}(\rho u \vec{u}) = \mathrm{div}(\mu \mathrm{grad} u) - \frac{\partial p}{\partial x} + S_u \tag{4.23a}$$

$$\frac{\partial (\rho v)}{\partial t} + \mathrm{div}(\rho v \vec{u}) = \mathrm{div}(\mu \mathrm{grad} v) - \frac{\partial p}{\partial y} + S_v \tag{4.23b}$$

$$\frac{\partial (\rho w)}{\partial t} + \mathrm{div}(\rho w \vec{u}) = \mathrm{div}(\mu \mathrm{grad} w) - \frac{\partial p}{\partial z} + S_w \tag{4.23c}$$

p pressure on fluid microelements;
S_u, S_v, and S_w the general source term of the momentum conservation equation.

The Eq. (4.23) is the momentum conservation equation, referred to as the momentum equation, also known as the motion equation, the other name of which is the Navier-Stokes equation.

3. Energy conservation equation

The law of energy conservation is the basic law that is applicable to any flow system that contains heat exchange. The law can be expressed as: the increase rate of energy in the micro-unit is equal to the net heat flow into the micro-unit plus the work that physical strength and facial strength do to the micro-unit. This law is actually the first law of thermodynamics.

$$\frac{\partial (\rho T)}{\partial t} + \mathrm{div}(\rho \vec{u} T) = \mathrm{div}(\frac{k}{c_p} \mathrm{grad} T) + S_T \tag{4.24}$$

c_p the heat capacity;
T temperature;

k the heat capacity;

S_T the inner heat source of the fluid and the part of the fluid whose mechanical energy can be converted into heat energy by the viscous fluid.

Combining the basic equations (4.22), (4.23a), (4.23b), (4.23c), (4.23c), (4.24), it can be found that there are six unknowns namely u, v, w, p, T, and x, and a state equation that connects p and ρ are required to close the whole equations:

$$p = p(\rho, T) \tag{4.25}$$

By solving the above-mentioned fluid basic control equation, CFD makes a numerical simulation of complex flow field.

4.2.2 *Key Steps in Ansys CFX Flow Field Calculation*

CFX is a fluid analysis software owned by Ansys. Unlike most CFD software, CFX uses finite volume based on finite elements to absorb the numerical accuracy of finite element method. Its main calculation process and basic steps are displayed in Figs. 4.12.

4.2.2.1 Create a Geometric Model

Using CAD software, we can create geometric models so that they can clearly and concisely express the problems to be analyzed and make use of the symmetry of the models to simplify the calculations.

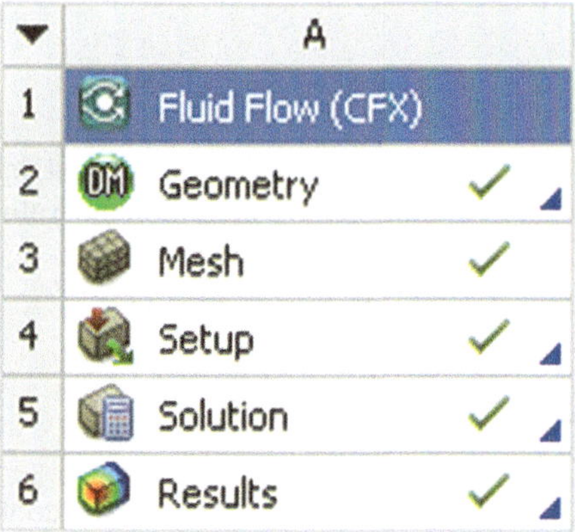

Fig. 4.12 Basic steps for Ansys CFX fluid field analysis

4.2.2.2 Divide the Computational Grid

When CFD analyzes the fluid field, it is necessary to use the grid to discrete the control equation in the spatial domain. Grids can be divided into structural grids and non-structural grids. Structure grids are spatially standardized, and grids are often distributed in rows. Rows and column lines are fairly obvious. For non-structural grids, there are no distinct rows and column lines in the spatial distribution.

Ansys CFX is equipped with a dedicated mesh generation tool, CFX-Mesh, which can divide the quadratic non-structural mesh and can also receive files generated by other software, such as the dedicated meshing software ICEM CFD.

4.2.2.3 Pre-processing (CFX-Pre)

Pre-processing is one of the most important aspects of Ansys CFX fluid field calculation, which includes determining fluid domain, setting material characteristics, setting boundary conditions, selecting solvers and operating environment, adjusting the relevant parameters for controlling solutions, and initializing the fluid field, etc.

For conjugate calculation of fluid and temperature fields, two solution domains (Domains) need to be set up, one for heat ingress and one for cooling water. It's also necessary to set up subdomains (Subdomains) on the winding model to load the heat source.

4.2.2.4 Solution (CFX-Solver)

By observing the change rate of the related variables in the process of solving, the convergence and stability of the solution can be monitored. These variables include turbulence variables such as velocity, pressure, temperature, and kinetic energy.

4.2.2.5 Post-processing (CFD-Post)

The results of the solution are displayed, including a fluid chart, a distribution cloud picture of temperature, pressure, and velocity, and a curve of change of transient simulation variables over time.

4.2.3 Motor Cooling

According to the cooling medium, the cooling mode of the motor is mainly divided into air cooling, hydrogen cooling, water (oil) cooling, evaporation cooling and superconducting motor, which can be further divided into internal and external cooling

Table 4.5 Thermal loads for different cooling modes

Cooling	Air-cooled	Hydrogen external cooling	Hydrogen internal cooling	Water cooling
Stator Electro-density J_1(A/mm^2)	2.5–3.5	3.5–4.5	6–8	8–12
Line load AS (A/cm)	500–600	600–800	1000–1300	1800–2400
AS·J_1	1600–2000	2100–3600	500–9000	12,000–25,000
Rotor electro-secret J_2 (A/mm^2)	3–5	4–4.5	5–8	8–12
Stator thermal load q_1 (W/cm^2)	0.4–05	0.7–1	1.2–2.4	3–5

according to the position of cooling mode relative to the heat source. The thermal loads in several cooling modes are shown in Table 4.5.

4.2.3.1 Air-Cooled

Air cooling is safe and reliable without risk of corrosion and is conductive for budget. However, it can be seen from the above table that the cooling effect of the motor is poor and the air volume loss is large. This will be detrimental to the efficiency of the high-speed running pulsed alternator.

4.2.3.2 Hydrogen Internal Cooling

Clamp a number of air-core stainless steel tubes into the solid core copper wire or making the windings with air-core copper wire can lead the hydrogen fluid through and export the heat outside, this is called hydrogen internal cooling. Because of the large density and small thermal conductivity, the fluid resistance of hydrogen is rather small, which has a decent cooling effect. Under the same pressure, the ventilation loss of hydrogen cooling and wind loss are 1/10 of the air respectively, which improves the cooling efficiency of the motor, and is a cooling method which is widely used.

4.2.3.3 Water Cooling

The heat capacity and thermal conductivity of liquid is larger than gas, so the cooling capacity of liquid cooling is greatly improved compared to air cooling. Water is an ideal cooling medium, it has a great ratio of heat and thermal conductivity, allowing more electromagnetic load than air-cooling and hydrogen cooling. The internal cooling mode tends to directly act on the heating parts with a very significant cooling effect. This is an ideal cooling way for air-core pulsed alternator.

4.2.3.4 Evaporation Cooling

The evaporation cooling is to use the principle of liquid vaporization and heat absorption of the cooling medium to cool the motor. Evaporation cooling is the most efficient cooling method theoretically because of the large vaporized heat and small required fluid. The temperature difference between the parts of the winding is so small that it has become a new direction of the current cooling technology. But at present, evaporation cooling is still in the stage of trials. There are still some technical problems and it lacks long-term operation test.

4.2.3.5 Superconductive Motors

Unlike the preceding cooling methods, the superconducting motor essentially solves the heat problem in the motor by reducing the motor loss. The motor with superconducting technology has a series of advantages including small size, light weight, low loss and high efficiency. With the development of the second generation of high temperature superconducting strip YBCO, the manufacturing process and operating conditions of high temperature superconducting motor are gradually reduced, the cost control is more reasonable, with good prospects for development.

Due to the demand of the high energy density to the motor, in order to reduce wind loss, it's not favorable for the high-speed air-core pulsed alternator to use air and hydrogen cooling. If necessary, a vacuum treatment is needed; The main challenges for evaporation cooling and superconductive motors are the cost and the maturity of technology. It is the ideal cooling method for pulse alternators and can be used as the preferred cooling solution.

4.3 Cooling Design of Pulsed Alternators

From the analysis of Sect. 4.1.4, it is obvious that the extremely low thermal conductivity of composite materials and the natural convective cooling structure of the model prototype greatly limit the number of discharges and discharge frequency of the air-core CPA. From the temperature distribution map it can also be clearly seen that the temperature of the heat source winding is significantly higher than other components, the insulation of composite materials seriously affected the heat dissipation of the heat source winding. Therefore, air cooling and water cooling which are commonly used in small and medium-sized induction motors and permanent magnetic synchronous motor are not suitable for the air-core CPA. If continuous discharge needs to be achieved during operation, the heating winding of the air-core

CPA must be cooled directly. This section refers to the water-cooling structure of large steamed turbine alternator, designs two kinds of active cooling structure for the model prototype, and carries on the simulation analysis. Both cooling structures are designed for sub-armature windings with the highest temperature rise, assuming a working period to be 30 s.

4.3.1 Active Cooling Structure 1

The cooling structure 1 model, as shown in Fig. 4.13, when making the sub-armature winding, six stainless steel pipes with an outer diameter of 6 mm and an inner diameter of 5 mm each is evenly placed inside the winding. Because the model winding is not that long, the series waterway is used here to meet the cooling requirements and simplify the process structure.

In the simulation calculation, the winding model is treated with equivalent processing, the copper wire, insulating paint and epoxy resin are equaled to a thermal conductor, and the equivalent winding is set to be in the solid solution domain. The thickness of the stainlesssteel tube is ignored here, and the cooling water in the tube is set to be in the fluid solution domain. Because the temperature is required, both solution domains need to activate the Energy Conservation Equation (Thermal Energy). Importing boundary condition at the entrance of the water pipe is set to be a constant current rate of 2 m/s with the temperature of inlet water being 40 °C. The exiting boundary condition at the entrance of the water pipe is set to be a constant pressure of 0 Pa, and the other boundary faces in the model, i.e. fixed wall boundary conditions, are defaulted. The standard k-epsilon model is selected for the turbulence model.

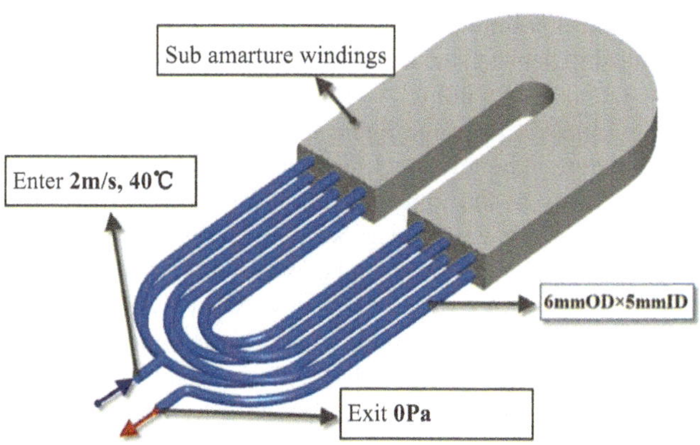

Fig. 4.13 Diagram of cooling structure 1

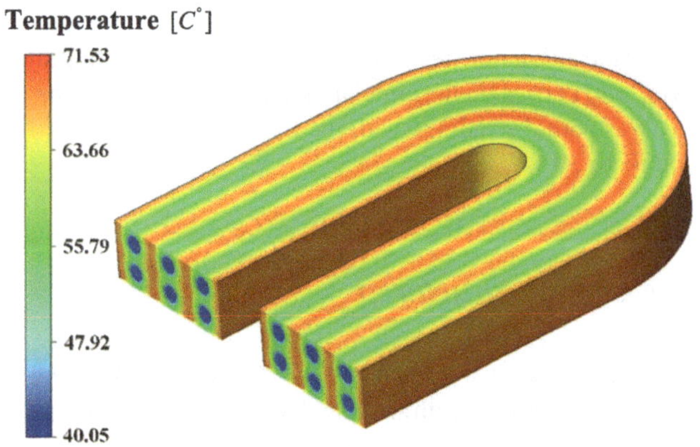

Fig. 4.14 Steady-state winding temperature distribution cloud map under cooling structure 1

The winding temperature distribution in a steady state is shown in Fig. 4.14, and the temperature distribution of the winding cross-section is uniform. The hottest spot is on the surface of the winding, at a temperature of 71.53 °C, which is below the limited temperature of epoxy resin and leaves a certain amount of domain. It shows that under this cooling structure, the sub-armature winding steam can be continuously discharged in a discharge cycle of 30 s.

Using the post-processing function of Ansys CFX software, some other conclusions can be drawn: after cooling the winding through the heating winding, the water temperature along the course increases continuously, and the water temperature rises from 40 °C at the inlet to 41.56 °C; the rate of water flow is also increased from 2 m/s at the inlet to 2.46 m/s; the entire cooling tube pressure is reduced to 48 kPa, and the average convective heat exchange coefficient for the relative winding is 11,680 $W/m^2 \cdot K$.

The cooling structure 1 is simulated and analyzed by the variable of cooling water flow, and the relationship between the temperature of the hottest spot on the winding and the water flow rate is shown in Fig. 4.15.

It can be clearly objected from the figure that the water flow rate has a direct impact on the temperature of hottest spot of the winding. The decline of the winding temperature takes 1.5 m/s as the inflection point and slowed down significantly afterwards. When the water flow rate exceeds 2 m/s, its impact on the hottest spot is basically unchanged. This phenomenon can be explained from two points of views. First of all, as the flow rate increases, the cooling water heats in the tube for shorter periods, and the temperature of the winding decreases as the temperature of the winding increases. Secondly, based on the basic theory of fluid dynamics, when the flow rate is low, the number of Reynolds in the tube is less than the number of the lower critical Reynolds, the flow pattern of the tube is in layers, the convection heat coefficient of the relative winding is small, and the winding temperature is higher. When the flow rate continues to rise to a certain value, for example, 1.5 m/s, the Reynolds number is 15,800, which is greater than the upper critical Reynolds number

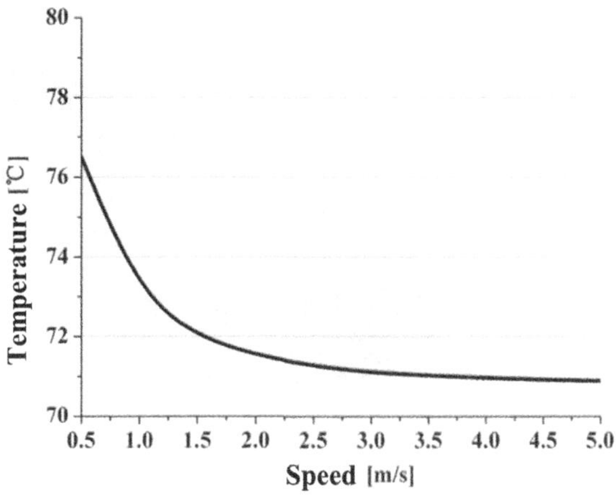

Fig. 4.15 The relationship between the temperature of the hottest spot on the winding and water flow rate in cooling structure 1

13,800. The fluid comes into turbulence state, and the convection heat exchange coefficient increases dramatically. The temperature of the hottest spot on the winding significantly decreased. When the flow keeps accelerating, due to the complete form of the fluid, it doesn't play a decisive role in the windings anymore, which helps the hottest spot to be in a steady state. Therefore, this section assumes the flow rate of the inlet water in cooling structure 1 is 2 m/s. It can be ensured that the winding temperature is maintained at a low value, without giving extra burden to the pump.

4.3.2 Active Cooling Structure 2

The schematic of cooling structure 2 is as shown in Fig. 4.16, in which the painted pack wire winding sub-armature winding is replaced by the thin air-core copper pipe with a 3 mm outer diameter and a 1.5 mm inner diameter. This is similar to large steamed turbine generator where de-ion water flows through the air-core copper pipe, taking away the heat generated by the winding group during the process of self-excitation. The number of winding turns was reduced from 250 to 32. In order to maintain the magnetic potential, the effective value of the current flowing through the copper pipe was raised from 221 A to 1727 A, and the corresponding winding heat rate was also increased to 6.096×10^6 W/m^3.

The steady-state conjugate CFD simulation analysis is carried out on the basis of cooling structure 2, the air-core copper conductor is set to be in the solid solution domain and the in-tube cooling water is set to be in the fluid solution domain. The boundary conditions are the same as the cooling structure 1 setting: inlet flow rate is 2 m/s; water temperature is 40 °C; outlet water pressure is 0 Pa; the turbulence model k-epsilon is in use.

Special treatment is required when meshing a simulation model because the fluid model is a thin and long pipe with a small radial length (only 1.5 mm in diameter) and

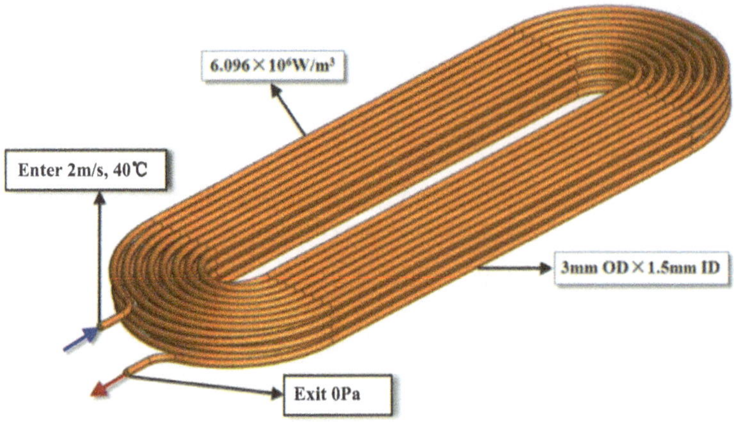

Fig. 4.16 Diagram of cooling structure 2

axial length (approximately 16 m). Because the long-diameter ratio of non-structured tetrahedron mesh increases, the mesh quality decreases remarkedly and the resulting solution is inaccurate. It's therefore necessary to increase the number of nodes and units, so that the time required for sectioning, computing and the computer memory is greatly increased. If the hexagonal section of the structural mesh is used, the distance between the shaft upward nodes can be increased appropriately, and the diameter up to the fluid boundary layer can be properly refined, ensuring the calculation accuracy, and at the same time saving the calculation and sectioning time. The diagram of the two methods of sectioning are shown in Fig. 4.17.

Comparing the above two figures, it can be seen that the number of nodes are reduced appropriately because the physical quantities such as water warm water pressure do not change much in the axial direction. With hexagonal mesh, the number of nodes is cut axially, and the boundary layer is encrypted radially, resulting in less number of units and number of nodes. Moreover, from the principle of computational fluid dynamics, the iterative equation is simpler and the calculation is more accurate by using the unit of the structural grid.

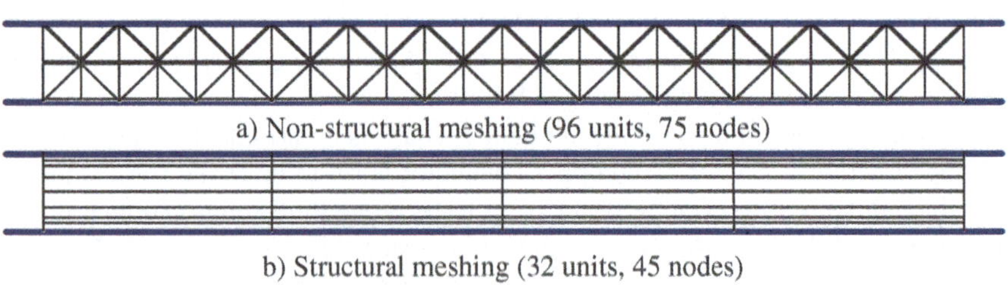

Fig. 4.17 Comparison of structural mesh with non-structural mesh

Fig. 4.18 Grid division results

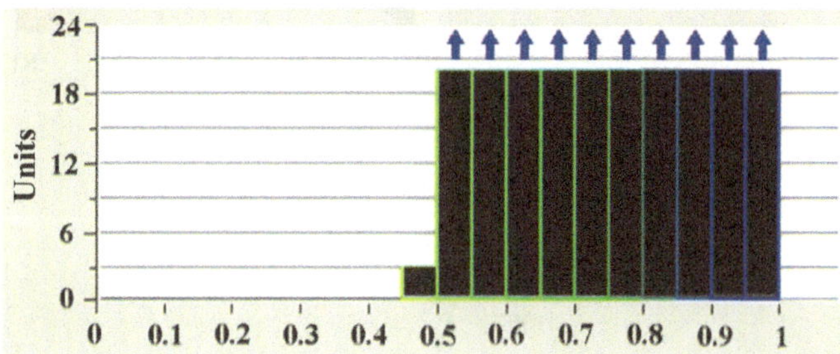

Fig. 4.19 Grid quality

In summary, this section uses professional meshing software Ansys ICEM CFD for a hexahedral structure meshing within the cooling structure 2. The corresponding grid division results as shown in Fig. 4.18, with 940,652 units and 813,566 nodes. The grid quality evaluated by the software is shown in Fig. 4.19, and it can be seen that the grid quality of each unit is greater than 0.4 (it is generally believed that the grid quality is greater than 0.3, that is, the calculation requirements of composite engineering).

The temperature distribution of the steady-state winding of the cooling structure 2 is shown in Fig. 4.20. At the outlet of the cooling water, temperature of the winding rises to 81.99 °C, which fulfills the requirements of the cooling design and enable continuous discharge operation. After post-processing these can also be calculated: water flow rate at the outlet is 2.31 m/s; the entire cooling tube pressure is reduced to 0.9 MPa; the average convection heat exchange coefficient for windings is 14,750 W/(m^2·K).

Similarly, the cooling structure 2 is simulated and analyzed as a variable, and the relationship between the hot spot temperature of the winding and the water flow rate is shown in Fig. 4.21.

As can be seen from the figure, for cooling structure 2, the water flow rate has a greater impact on the hottest spot temperature of the winding, mainly because the waterway of cooling structure 2 is longer, in this way the water flow rate directly affects the time of the cooling water to run through the heated winding. As the flow rate increases, the heating time of the cooling water along the course is shorter, the

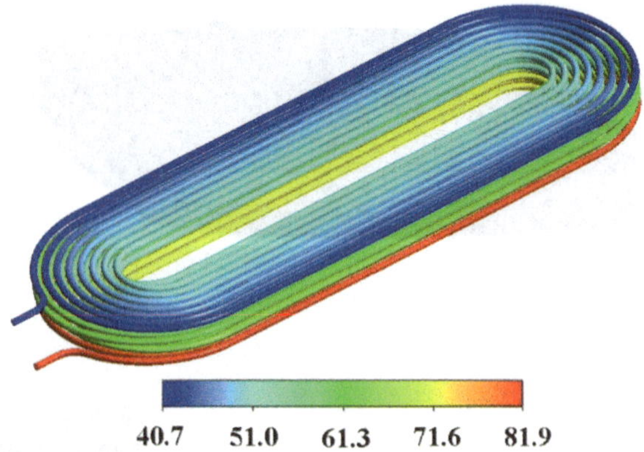

Fig. 4.20 Steady-state winding distribution cloud picture in cooling structure 2

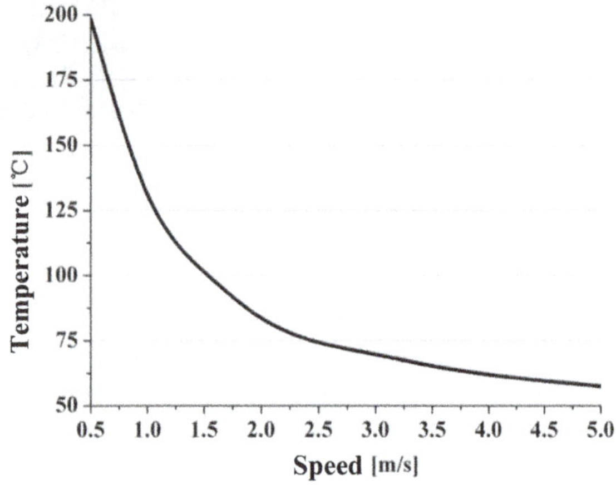

Fig. 4.21 The relationship between the temperature of the hottest spot on the windings and water flow velocity

temperature rise decreases, and the temperature of the hottest spot of the winding is significantly reduced. Based on the above analysis, the inlet water flow rate is set to be 2 m/s under cooling structure 2, to ensure that it works below the limit temperature, without speeding up the wear of copper pipe due to high flow rate.

4.3.3 Comparison of the Two Cooling Structures

From the simulation results, it can be seen that when the discharge cycle is 30 s in length, both cooling structures meet the goal of continuous discharge operation of the sub-armature winding. However, cooling structure 1 does not change the original winding structure, only to make it simpler and easier to implement. While the stainless steel tube only allows water to pass through rather than electricity, so that the insulation problem is not necessary to be considered. However, the addition of stainless steel tube and cooling water on the original winding of structure 1 enlarges the winding's volume and quality. Through calculation, the volume of the winding which was cooled by intubated stainless steel tube increased by 17.7%, the mass of which increased by 13%. Cooling structure 2 adopts air-core copper pipe instead of fine painted wire pack, which reduced the number of the turns of the winding and increased the sectional area, so that the inductance and resistance of sub-armature winding are significantly reduced. It's also more conducive to the establishment of self-stimulation process. Furthermore, the mass of the cooling structure 2 diminished remarkedly. The coil quality of a single sub-armature winding was reduced from 2.48 kg to 1.35 kg. A decrease of up to 45.5% is very meaningful for increasing the power density of the pulsed alternator However, because the air-core copper pipe of structure 2 is both water-based and conductive, the system needs an extra deionization device. In addition, the winding and manufacturing of air-core copper pipe are more complex. It also needs real-time monitoring of the conductivity of cooling water during operation. If it's in a long term, the scale and other issues need to be taken into consideration.

In summary, the two cooling structures have their own advantages and disadvantages. This chapter only put forward the design of the cooling system and focused on the simulation computational methods. The analysis of specific design and manufacturing process depends on the actual problems.

Chapter 5
Mechanical Performance of Pulsed Alternators

Compensated Pulsed Alternators (CPAs) typically operate at high speed to obtain higher inertial energy storage and reverse potential for higher energy storage and power densities. CPA is subjected to extremely high mechanical stress. Moreover, during pulsed discharging, the excitation current and discharge current are very high, there exists a very strong magnetic field, in the motor instantaneously. Some components of the motor are therefore subjected to a great instantaneous electromagnetic force, which directly threatens the security of CPA, so the mechanical stress, electromagnetic stress and key component strength are required to be analyzed and verified. The introduction of fiber resin composites with high strength-density ratio decreases the quality and increases the speed of the air core CPA, which greatly improves the energy density and power density. However, due to the anisotropy characteristics of the mechanical characteristics of fiber resin composite materials, the lateral strength of fiber is much lower than the vertical strength. It is necessary to carry out a safety assessment for reasonable parameters design.

5.1 Analysis of the Mechanical Performance of Pulsed Alternators

As a rotating machine with inertial storage, the mechanical performance analysis for CPA includes the analysis to mechanical stress and electromagnetic stress. In order to obtain the general law of CPA force analysis, it is necessary to dig further in the source and computational method of mechanical stress and electromagnetic stress. The mechanical research of CPA is mainly based on safety verification in the initial design. For a better understanding of the vibration and noise of CPA, not only the total force and torque of the motor need to be calculated and verified, but also the distribution of the internal force density of the motor needs to be considered for optimization.

S. Wu and S. Cui, *Pulsed Alternators Technologies and Application*,
https://doi.org/10.1007/978-981-33-4224-8_5

The mechanical stress analysis and the theoretical study of the strength of the motor mainly aim at large-scale hydro-generators and turbo-generators. The analysis of allowed mechanical stress of the motor mainly includes the verification of motor strength, and the application of allowed stress analysis and material mechanics in motor design. Although the CPA is smaller in size than large hydro-generators and turbo-generators, the pulsed power peak is comparable to the rated power of large generators due to the high-speed operation of CPA, and the instantaneous impact of the components is equally large, requiring stress analysis and strength verification of its internal parts. The mechanical performances of the rotor of iron core CPA, including the fit tolerance of the iron core, the calculation of the separate speed of the motor and the critical speed of the rotor, are similar to the conventional iron core motor. They are both common problems in the analysis of mechanical performances. Instead of these mentioned above, this book merely analyzes the mechanical performances of air core CPA.

5.1.1 Mechanical Stress of Pulsed Alternators

The high-speed rotating rotor of pulsed alternator is subject to great centrifugal force, which may cause stratification of rotor yoke, and compensate for the severe consequences of the deformation of pulsed alternator's compensation shield and epoxy cracking. Therefore, it is necessary to analyze the rotor strength under the influence of centrifugal force. In field of material mechanics, strength usually refers to the fracture or plastic deformation occurred in the material when the part is subjected to the internal force which is beyond the yield limit. The definition of "strength" of motor structure is different. It refers to the deformation of the motor structure under forces when the motor is working. Even if it is still in the elastic range, there is no break or plastic deformation in the structure itself, but the motor has not been able to work properly. It is also defined as the strength failure of the motor structure. The standard of strength failure of high-risk components in the pulsed alternator is defined as below.

5.1.1.1 Strength of the Compensation Shield

In the air core CPA, the compensation shield is glued to the outside of the rotor by epoxy resin through the intermediate bonding layer to transmit the torque between the two. For example, high-speed rotation enlarges the internal stress of the bonding layer to the yield limit of the resin, and the compensation shield disengages from the rotor like the iron core machine. At this time, the rotor and the motor cannot work properly. Although in both cases, the internal stress of the compensation shield may be less than the yield limit of its material and there is no fracture in the tube, the motor structure has already failed because it cannot work properly anymore or fearing dangerous occurrence. Therefore, the strength of some structural parts of the motor is different from the concept of the strength of the material in material mechanics, but the deformation of the structural parts is limited to a certain range.

5.1.1.2 Strength of the Air Core Rotor Yoke

In the air core CPA, the yoke of the stator and rotor of the motor is a cylinder formed by fiber epoxy composite. Fiber is a material of strength anisotropy with enormous vertical strength and small transverse strength. Through force analysis of the composite material winded in the direction of the ring, it can be seen that if the thickness of the composite material reaches a certain degree, the radial stress during high-speed rotation will exceed the radial strength of the fiber and result in stratification. Although the whole composite yoke has not broken, it is not possible to transmit the torque in a desirable way because of the stratification, which results in the failure or dangerous occurrence of the rotor.

5.1.1.3 Strength of the Epoxy Banding Layer

It's difficult for the carbon fiber epoxy composite material to withstand the high temperature required by the shrink fit of the aluminum shield, and the high price and complex operation of the low temperature cooling equipment. The clearance fit is hence adopted between the aluminum compensation shield and carbon fiber epoxy resin strap. The middle is coated with epoxy resin to increase the strength of bonding. The air core rotor is as shown in Fig. 5.4. When the assembled rotor is rotating, because the radius of the compensation shield is slightly larger than the radius of the rotor, the centrifugal force on the compensation shield is more than the entire rotor, and the deformation of the entire rotor's outer diameter is relatively small compared to the inner diameter of the tube. The higher the speed is, the greater the difference between the two deformations will be. If the stress caused by deformation is within the elastic range of the epoxy resin, the parts will not be separated from each other, the rotors will not be malfunctional and the CPA can continue to discharge. If the stress in the epoxy layer exceeds its allowed stress limit, the compensation shield and the entire rotor are out of contact and the rotor fails.

5.1.2 Electromagnetic Stress of Pulsed Alternators

When the current flows through the armature, the force that the armature winding is subjected to is called electromagnetic force under the influence of the magnetic field within the motor. During CPA discharge, the current within the armature winding changes, and usually when the winding current reaches its peak value, the pressure on the winding is also at its maximum. The greater electromagnetic stress places higher demands on the strength between the winding bandage, the stator bandage and the winding and the composite yoke.

5.2 Research Methods for Stress of Pulsed Alternators

5.2.1 Research Methods for Mechanical Stress

5.2.1.1 Analytical Method

The main mechanical problems of the high-speed rotation of the core CPA include the radial stratification problems between different composite material layers or the composite materials and compensation shields due to massive centrifugal force under high speed. This section focuses on the computational method of stress between air core CPA rotor layers in order to check whether the stratification problems occur during the high-speed rotation of the rotor.

The rotor of the air core CPA can be regarded as a set of multiple tubes. Based on axial symmetry, it can be equivalent to two-dimensional plane model for analysis. Given the basic theory of thick-wall tube in elastic mechanics, this section takes a space-rotating tube as an example to clarify the analysis of mechanical stress.

For a micro-unit in the tube, the force analysis is carried out, as shown in Fig. 5.1. The inner diameter is R_1, the outer diameter is R_2, the thickness is h, the radial length is dr, and the circumferential length is $d\varphi$.

Assuming that the disc is not subject to axial stress and it rotates at a high speed, only radial strain occurs. There is no internal shear stress or shear strain. It is only affected by radial positive stress σ_r and circumferential positive stress σ_φ, both of which are functions of r. It is gathered that the radial positive stress σ_r and the circumferential positive stress σ_φ are:

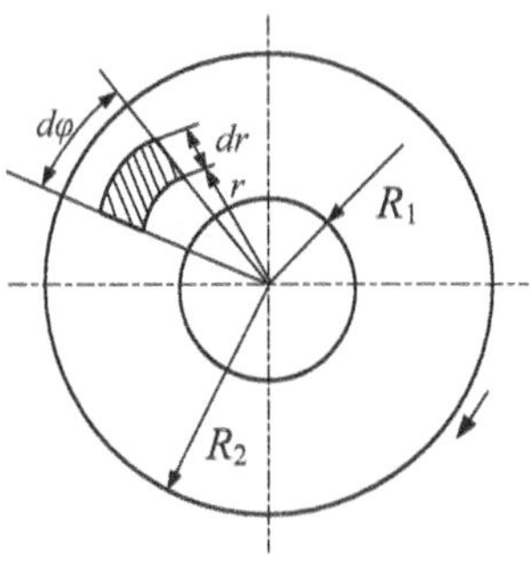

Fig. 5.1 Force on the microstructure in the tube

$$\begin{cases} \sigma_r = \dfrac{E}{1-\mu^2}\left[(1+\mu)C_1 - (1-\mu)\dfrac{C_2}{r^2} - \dfrac{1}{8}(3+\mu)\left(1-\mu^2\right)\dfrac{\rho\omega^2 r^2}{gE}\right] \\ \sigma_\varphi = \dfrac{E}{1-\mu^2}\left[(1+\mu)C_1 + (1-\mu)\dfrac{C_2}{r^2} - \dfrac{1}{8}(3\mu+1)\left(1-\mu^2\right)\dfrac{\rho\omega^2 r^2}{gE}\right] \end{cases} \tag{5.1}$$

E—elastic modulus of the material;

μ—Poisson ratio;

ρ—mass density;

ω—rotational angular velocity;

C_1, C_2—integral constant, according to the boundary conditions of the inner and outer edges of the tube.

Based on Formula (5.1), when the high-speed rotating tube is given different material properties and different boundary conditions, the stress distribution at any position within the tube can be obtained. However, the equation is for isotropic materials. For the anisotropic rotor yoke, the equation needs to be adjusted to some degree, although the core idea of the analytical solution method is based on formula.

5.2.1.2 Finite Element Method

The finite element method of structural analysis was founded by a group of researchers from industry and academia in the 1950s and 1960s. The main steps to apply finite element method in structural analysis include: structural dispersion, which is to divide the original structure artificially into several sub-regions (units), which are connected to each other by finite nodes; selection of the displacement interpolating functions, which represents the displacement of any point within the cell as the interpolation of the node displacement, and organizes it into a matrix; cell analysis, which means calculation of cell stiffness matrix and equivalent node load vector; overall analysis and constraint processing, that is forming cell stiffness matrix and load vector set, introducing constraints, and eliminating rigid body displacement to establish structurally near equilibrium differential equations, therefore solving the equilibrium equations and calculate cell stress. Among them, structural dispersion and constraint processing are carried out through analyzing the thinking of the personnel and manual operation. The quality of the grid and whether the constraints directly affect the final analysis and calculation results and the remaining steps can be done automatically through the powerful functions of large-scale general software. The two tasks of dispersion and the simplification of the analyzed structure take 80–90% of the whole process of the total finite element analysis.

At present, the large-scale universal finite element software which are commonly used in structural analysis include ADINA, ABAQUS, ANSYS, MSC/Marc, MSC/Nastran, etc., this paper takes ANSYS/Workbench as an example to introduce

Fig. 5.2 Basic steps for the analysis of an Ansys Workbench structure

the process of applying finite element software to solve mechanical stress problems (Fig. 5.2).

The simulation analysis process includes: (1) geometric model and material properties, (2) type of contact, (3) analysis settings, (4) environment, such as load and constraint, (5) solution model, and (6) results and post-processing.

(1) Model and material properties

The model includes the establishment of the geometric model and the division of the finite element model. The establishment of geometric model is based on the function of finite element software or 3D drawing software. In view of the complex rotor structure in this book, when establishing the geometric model, the selection of the corresponding geometric model should be made according to the different effects to the target performance of the whole structure of each part. The shape of the components should be simplified accordingly, taking into account the contents of the analysis. Material properties include defining the material parameters as elastic modulus, Poisson ratio, density, and thermal expansion coefficient of each component.

(2) Type of contact

The types of contact between the mating surfaces are binding, non-separation, frictionlessness, roughness, and frictional contact. The contact settings are carried out in accordance with the cooperation relationship between the components.

(3) Analysis settings

The analysis setting provides general solution process control. The step control is about manual time step control and automatic time step control, specifying the number of analytical steps in analysis and the termination time of each step. Time in static analysis is treated as a tracking mechanism. The solution control is about direct solution and iterative solution.

(4) Environment

Loads and constraints are added to the analysis model, which are defined in the form of the degrees of freedom of the selected unit. The types of loads are: inertial loads, which are applied to the entire model, and the material density needs to be calculated, which refers to the forces applied to the defined mass point; structural loads, as in the force or torque applied to the components of the system; structural constraints, which are used to avid the movement of certain regions; thermal loads, which creates a temperature field and expands or conducts the heat in the model.

(5) Solution of the model

It refers to the solution of the model.

(6) Results and post-processing

Checking and post-processing of the results. There could be many different results of the post processing: deformation in all directions and total deformation, stress–strain components, main stress–strain or stress–strain invariant.

5.2.2 Research Methods for Electromagnetic Stress

5.2.2.1 Analytical Method

As an electromechanical energy conversion component, CPA relies on electromagnetic force and electromagnetic torques for electromechanical energy conversion. The calculation theory of electromagnetic force is the basis of electromagnetic force analysis. According to the first law of thermodynamics, for a system, between two different states, the change in energy in the system is equal to the work done by electromagnetic force, the joule heat of the system, and the sum of the energy applied by the outside world to the system through the surface.

$$-dW = P_j dt + P_s dt + dL \tag{5.2}$$

P_j—power of joule loss $P_j = \iiint_V \overline{E} \cdot \overline{J} dV$;

P_s—external energy provided to the inside of the system;

$P_s = -\oiint_{S_v} \left(\overline{E} \times \overline{H}\right) \cdot d\bar{s}$;

dL—the work done by electromagnetic forces.

In order to quickly find out the work done by electromagnetic force, it is assumed that $P_j dt + P_s dt = 0$. From the view of power, it can be displayed as following after analyzing Formula (5.2):

$$\frac{dL}{dt} = \iiint\limits_V \bar{f} \cdot \bar{u} dV = -\frac{d}{dt}\left(\iiint\limits_V w_e dV\right)\Bigg|_{\Psi e=const} - \frac{d}{dt}\left(\iiint\limits_V w_m dV\right)\Bigg|_{\Phi m=const} \tag{5.3}$$

f—unit physical density;

u—the speed of the unit;

J—current density;

D—electric displacement;

E—electric field strength;

H—magnetic field strength;

B—flux density;

ε—dielectric constant;

μ—magnetic conductivity;

w_e—electrical energy density $w_e = \frac{\overline{D} \cdot \overline{E}}{2} = \frac{\varepsilon E^2}{2} = \frac{D^2}{2\varepsilon}$;

w_m—magnetic energy density, $w_m = \frac{\overline{B} \cdot \overline{H}}{2} = \frac{\mu H^2}{2} = \frac{B^2}{2\mu}$.

With the help of Maxwell equations $div\overline{D} = \rho_0$ and $div\overline{B} = 0$, and the conditions on the selected boundary planes, the right side of Eq. (5.3) is simplified to

$$\iiint\limits_V \bar{f} \cdot \bar{u} dV = \iiint\limits_V \left[\rho_0 \overline{E} - \frac{E^2}{2} \cdot \nabla\varepsilon + (\overline{J} \times \overline{B}) - \frac{H^2}{2} \cdot \nabla\mu\right] \cdot \bar{u} dV \tag{5.4}$$

It can be seen that the formula for calculating body density f is:

$$\bar{f} = \rho_0 \overline{E} - \frac{E^2}{2} \cdot \nabla\varepsilon + (\overline{J} \times \overline{B}) - \frac{H^2}{2} \cdot \nabla\mu \tag{5.5}$$

5.2.2.2 Finite Element Method

The method of calculating magnetic field is mainly finite element method, including differential method and Galerkin method (one of the weighted residual methods). The electromagnetic field owns high convergence speed, and the result of electromagnetic field can be solved quickly and accurately by finite element method. However, the low speed of convergence speed of force makes it more difficult to solve the force quickly and accurately from the finite meta-field solution. Therefore, the spatial distribution of electromagnetic field is solved by finite meta-field solution, and the spatial distribution of electromagnetic force is solved according to Formula (5.5).

In CPA, the formula for solving magnetic forces with electromagnetic field solutions is shown in Eq. (5.6). The first item on the right side of the equal sign becomes Lorenz force, which is related to current and magnetic field. The second is called surface force, which is related to the change gradient of the magnetic conductivity of the internal material of the CPA.

$$\overline{f_m} = \overline{J} \times \overline{B} - \frac{1}{2} H^2 \nabla \mu \tag{5.6}$$

Stress distribution will have an effect on the magnetic conduction of ferromagnetic materials, which in turn will affect the distribution of magnetic fields and the loss of iron cores of motors. However, for the empty core prototype with no internal non-magnetic material inside, when discharging, stress barely has effect on the material's magnetic permeability, so the one-way coupling of the electromagnetic structure can be used to calculate the magnetic field and stress deformation in the motor. That is, firstly, by calculating the electromagnetic field, the distribution of electromagnetic force in the motor is obtained, and then the electromagnetic force is replaced into the structural analysis model as the load of structural analysis. In this way the stress distribution and deformation in the motor can be gained. The one-way coupling analysis flowchart of electromagnetic structures is shown in Fig. 5.3.

The coupling analysis of electromagnetism and structure is carried out by finite element software. First, the electromagnetic analysis is carried out by the finite element software, and then the value of electromagnetic node force is exported to the Nastran files to analyze and calculate the stress and strain.

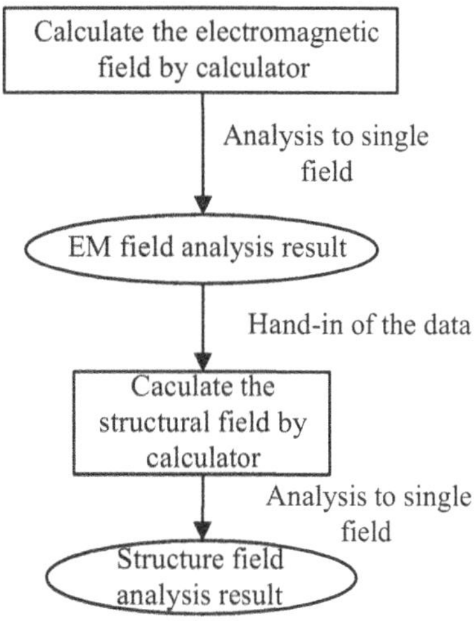

Fig. 5.3 One-way coupling analysis flowchart for electromagnetic structures

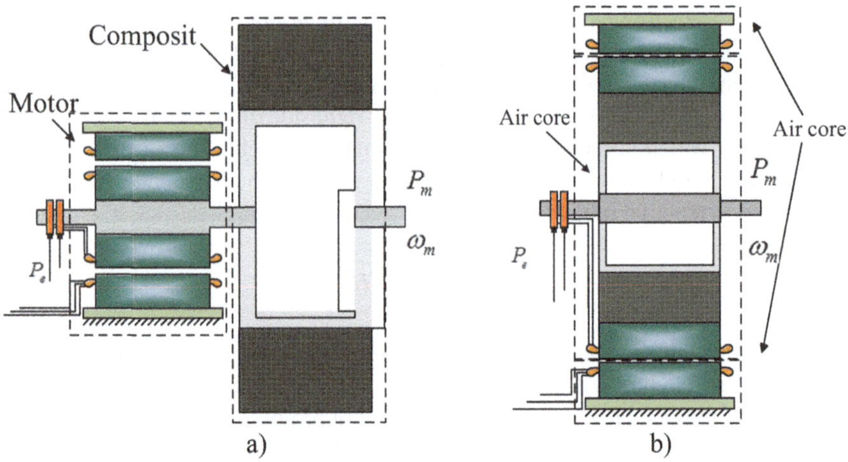

Fig. 5.4 Structural diagram of the traditional flywheel energy storage motor (**a**) and the air core pulsed alternator (**b**)

5.3 Stress Field Analysis of the Pulsed Alternators

The structural schematics of the traditional flywheel energy storage is shown in Fig. 5.4a, which consist mainly of a motor of highly efficient conversion, composite flywheels with the same yoke as the motor rotors, high-speed bearings and power electronic components required for power conversion. The schematics of the structure of the air core pulsed alternator with electro-excitation is shown in Fig. 5.4b, where the rotor includes a composite flywheel and a slotless winding. It can be seen that the compensated pulsed alternator can be regarded as a structure derived from the traditional flywheel energy storage structure, and its rotor force analysis is equivalent to the stress analysis of the composite flywheel rotor.

The difference, however, is that the electrical energy required by the pulsed load is converted from the inertial energy stored when the rotor rotates at high speed with electromagnetic induction and flux compression when the compensating pulsed alternator discharges. It produces a great electromagnetic stress in the windings and compensation. The following will be combined with the structure of the pulsed alternator in Fig. 5.4b to analyze its electromagnetic stress and mechanical stress.

5.3.1 Electromagnetic Stress Analysis

5.3.1.1 Force Analysis of Single Core Coils

For the air core CPA, because the magnetic conductivity of the composite structure of the fixed rotor is approximate to air, the gradient of the magnetic conductivity in

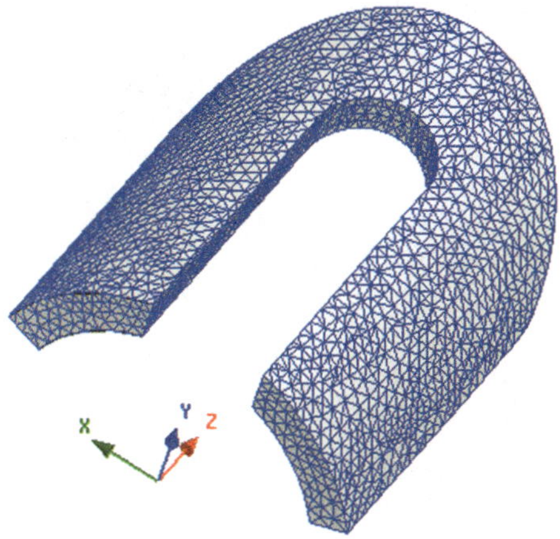

Fig. 5.5 3D finite element model of the winding

the space is approximately equal to zero, so from Formula (5.6), it can be seen that there is no face force density in the motor, and the physical density equals to the multiplication cross of the current density and magnetic density. Assuming there are no eddy currents inside the composite structure, the electromagnetic force in the air core CPA is all applied to the windings in which a current flow going through, and the body density of the Lorenz force on the windings is

$$\bar{f} = \overline{J} \times \overline{B} \tag{5.7}$$

In the air core CPA, the end leakage is large, and the end of the slotless winding is long. In order to take the force at the end into consideration, a three-dimensional force analysis needs to be carried out on the coils, so the formation of three-dimensional finite element model of the winding is necessary, as shown in Fig. 5.5.

The distribution of current density $\overline{J}$ and spatial magnetic field $\overline{B}$ within the winding is shown in Fig. 5.6. According to Formula (5.7), the vector and magnitude distribution of Lorenz force density are calculated, as shown in Fig. 5.7.

As can be seen from Fig. 5.7, for a single empty core coil, Lorenz volume force density is large at the end and the inside of the winding and the amplitude is pretty high, so in the design it's important to consider how to reduce the amplitude as of the force density of the motor winding can be reduced and evenly distributed. In this way the local stress concentration can also be lowered.

5.3.1.2 Air Core CPA Electromagnetic Stress Analysis

According to the finite element analysis method described in Sect. 5.2.2, the electromagnetic stress of the air core CPA is analyzed. The mechanical parameters of the core CPA internal material are shown in Table 5.1.

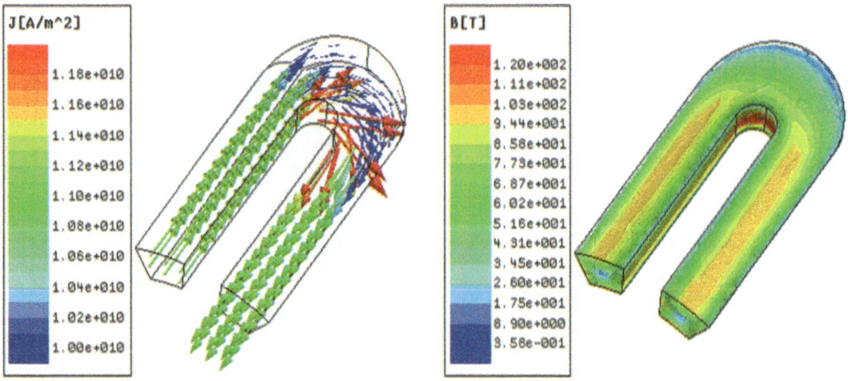

Fig. 5.6 Distribution of current density and magnetic field in windings

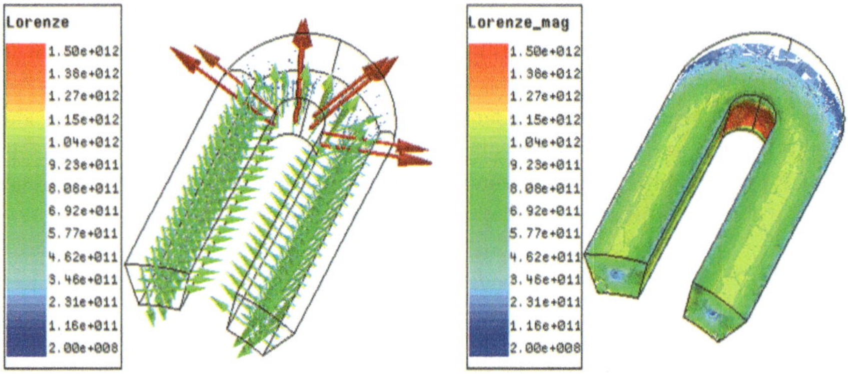

Fig. 5.7 Vectors and amplitudes of Lorenz force density distribution

Table 5.1 Mechanical parameters of the air core CPA internal material

Air core CPA internal material	Young modulus E (GPa)	Poisson ratio γ	Mass density ρ (kg/m^3)
Fixed rotor iron core	21	0.30	7850
Hard aluminum	69	0.33	2710
Red copper	110	0.33	8030
Carbon fiber	248	0.3	1800
S glass fiber	85	0.28	2480
Epoxy 3501-6	3.4	0.4	1200

Fig. 5.8 The distribution of the Lorenz force inside CPA prior to discharging

For hollow CPA, self-excitation or pulsed excitation are common. Prior to discharging, the excitation current is concentrated in the excitation winding, and the Lorenz stress inside the hollow CPA is concentrated on the excitation winding, as shown in Fig. 5.8.

In the process of discharging, there is a very significant instantaneous pulsed current flowing through the armature and the vortex is inducted in the aluminum compensation shield. At this time the distribution of the CPA Lorenz stress is as shown in Fig. 5.9.

When discharging, due to the compression of the compensation shield vortex to the armature reaction flux, the circumferential magnetic flux density accounts for the vast majority of the air gap, according to ampere law. At this time, the electromagnetic force on the main armature winding and the compensation shield is radial. As can be seen from Fig. 5.9, there is a great radial pressure effect on the armature winding and compensation shield during discharging. The discharging current is one or two levels higher than the excitation current, and the CPA internal parts are subjected to a large force during this process. As a result, the outside of the main armature winding is reinforced with a high-strength glass fiber epoxy resin.

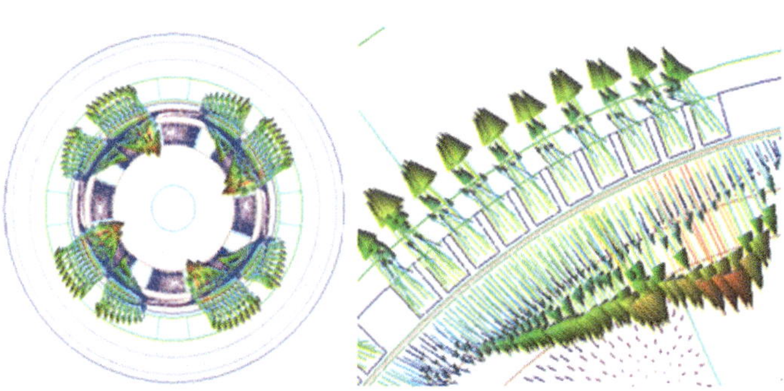

Fig. 5.9 Lorenz stress distribution at peak value of the discharging current

5.3.2 Mechanical Stress Analysis

5.3.2.1 Analysis of Mechanical Stress of Air Core Rotors

The mechanical stress analysis of the air core rotor focuses on the fiber resin flywheel and bandage. For composite flywheels, the stress in the fiber resin yoke and bandage can be calculated more precisely by finite element method, and then the strength core can be performed. However, the finite element method takes a long time, and due to the anisotropy of the composite material, it is usually necessary to establish a finite element model on the basis of testing. At the same time, finite element analysis is affected by mesh generation and is easy to disperse. Considering the simple cylinder-shaped structure of rotor flywheel and bandage, the analysis algorithm can obtain satisfactory results, so this paper mainly uses the analysis method to analyze the rotor's mechanical stress.

Due to the anisotropy of the fiber resin material, Formula (5.1) is adjusted to derive the stress analysis equation to meet the demand. In order to give full play to the advantages of high fiber axial strength, the composite flywheel is made of 90° hoop winding fiber The internal of single-layer composite flywheels are mainly affected by two forces during high speed rotation, which are the stress caused by centrifugal forces and the residual stress produced by resin curing. For long-straight composite flywheels, it is usually supposed that axial stress or strain is zero, and the inside of the flywheel is considered to be of a flat stress state, subjected only to circumferential stress and radial stress, and the stress analysis diagram of the volume element is shown in Fig. 5.10. r is the internal diameter of volume element, dr is radial length, $d\theta$ is the circumferential angle, l is axial length, σ_r is radial stress, σ_θ is circumferential stress, and σ_z is axial stress.

The equilibrium equation along the radial column is as follows,

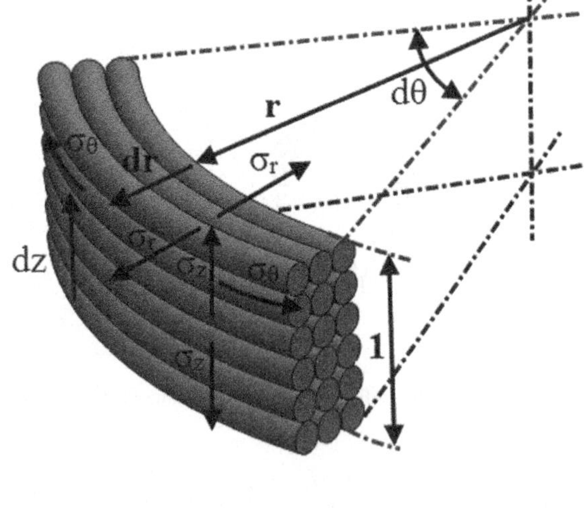

Fig. 5.10 Stress analysis diagram of composite flywheel volume elements

$$(\sigma_r + d\sigma_r)(r + dr)d\theta \cdot l - \sigma_r r d\theta \cdot l - 2\sigma_\theta dr \sin\frac{d\theta}{2} \cdot l = \rho r^2\omega^2 \quad (5.8)$$

Excluding the higher order infinitesimals, in the case of a very small $d\theta$, replace $d\theta/2)$ with $d\theta/2$ approximation. These can be derived from (5.8),

$$\frac{d\sigma_r}{dr} + \frac{\sigma_r - \sigma_\theta}{r} + \rho r\omega^2 = 0 \quad (5.9)$$

ρ—the density of the rotor material;

ω—the rotation angle speed of the rotor.

If the internal radial displacement of the volume element is u, and the outer radial displacement is $u + du$, and the following geometric equations can be listed as

$$\varepsilon_r = \frac{(u + du) - u}{dr} = \frac{du}{dr} \quad (5.10)$$

$$\varepsilon_\theta = \frac{(r + u)d\theta - rd\theta}{rd\theta} = \frac{u}{r} \quad (5.11)$$

ε_r—radial strain;

ε_θ—circumferential strain.

Differentiating r within Formula (5.11), and the latter can be simplified as:

$$\frac{d\varepsilon_\theta}{dr} = \frac{1}{r}(\varepsilon_r - \varepsilon_\theta) \quad (5.12)$$

Due to winding pull, resin viscosity and hardness, curing agent shrinkage and temperature expansion, the initial stress generated in the anisotropic composite material is also very large. Without appropriate parameters in the preparation process, the internal residual stress can easily result in layering and other failure in the composite tube. As shown in Table 5.2, the temperature expansion coefficients of fibers in axial and horizontal directions vary greatly, resulting in residual stress in the curing process, such as improper curing, which easily leads to excessive residual stress and layering of fiber resin tube. In this calculation, only the curing temperature that has a large impact on residual stress is considered, while other factors are ignored. The residual stress produced by temperature pressure drop ΔT during curing is given in the following relation between stress and strain according to the generalized Hook's law,

$$\begin{bmatrix} \sigma_r \\ \sigma_\theta \end{bmatrix} = \begin{bmatrix} Q_{rr} & Q_{r\theta} \\ Q_{r\theta} & Q_{\theta\theta} \end{bmatrix} \begin{bmatrix} \varepsilon_r - \alpha_r\Delta T \\ \varepsilon_\theta - \alpha_\theta\Delta T \end{bmatrix} \quad (5.13)$$

Table 5.2 Performance parameters for different fibers

Parameters	Glass fiber	Carbon fiber (T300)	Carbon fiber (T800)
Longitudinal tensile strength X (MPa)	1062	1800	2900
Vertical compression intensity X' (MPa)	610	1400	1600
Horizontal tensile strength Y (MPa)	31	80	70
Horizontal compression intensity Y' (MPa)	−118	−168	−168
Longitudinal modulus E_θ (GPa)	38.6	130	155
Horizontal modulus E_r (GPa)	8.27	9	9
Primary Poisson ratio $\nu_{r\theta}$	0.26	0.23	0.3
Poisson ratio ν_{rz}	0.3	0.3	0.3
Axial temperature coefficient α_θ (10^{-6}/°C)	8.6	−0.3	−0.3
Lateral temperature coefficient α_r (10^{-6}/°C)	22.1	28.1	28.1
Mass density ρ (kg/m^3)	1800	1600	1600

Q_{rr}, $Q_{r\theta}$, $Q_{\theta\theta}$—components of stiffness matrix Q, calculated by the elastic modulus of the fibers in all directions;

α_r—the radial temperature expansion coefficient of the fiber;

α_θ—the circumferential temperature expansion coefficient of the fiber.

In order to make the rotor lighter, the composite flywheel uses hollow hub support with a certain thickness. The modulus of the hub is greater than the carbon fiber, so at high speed, it's easy for them to disengage with each other. The united hub can be an effective solution. If the thickness t_s, the mass density is ρ_{h}, the pressure p_i that is generated on the composite flywheel on the inner diameter is

$$r = r_i, \quad \sigma_r = -p_i = \frac{\rho_h \omega^2 \left[r_i^3 - (r_i - t_s)^3\right]}{3r_i} \tag{5.14}$$

Due to the limitation of coil strength, the connection between the individual excitation coil and the composite flywheel does not provide sufficient conditions for interference fit to be realized. They are mainly bonded by epoxy resin. As long as a certain thickness is guaranteed, the stress inside the resin layer will not exceed the allowable stress. It can be hypothesized approximately that the radial pull of the resin layer to the outer diameter of the composite flywheel is zero.

$$r = r_o, \quad \sigma_r = 0 \tag{5.15}$$

The combination of (5.13) and (5.15) and the known stress at the inner and outer diameters of the composite flywheel are used as the boundary conditions for the solution, and radial and circumferential stresses inside the tube can be derived.

The circumferential stress σ_θ and radial stress σ_r produced by the single-layer composite ring under centrifugal force are as below respectively.

$$\sigma_\theta = \rho_{eff} v_0^2 \frac{3+\upsilon}{9-\mu^2}\left[\mu l\left(\frac{r}{r_o}\right)^{\mu-1} + \mu(l-1)\left(\frac{r}{r_o}\right)^{-\mu-1} - \frac{\mu^2+3\upsilon}{3+\upsilon}\left(\frac{r}{r_o}\right)^2\right] \tag{5.16}$$

$$\sigma_r = \rho_{eff} v_0^2 \frac{3+\upsilon}{9-\mu^2}\left[l\left(\frac{r}{r_o}\right)^{\mu-1} - (l-1)\left(\frac{r}{r_o}\right)^{-\mu-1} - \left(\frac{r}{r_o}\right)^2\right] \tag{5.17}$$

Moreover, $l = (\lambda^{-\mu-1} - \lambda^2)/(\lambda^{-\mu-1} - \lambda^{\mu-1})$; $\mu = \sqrt{E_\theta / E_r}$.

This is the ratio of the circumferential modulus to the radial modulus of the composite material, which reflects the anisotropy of the mechanical characteristics of the material.

Fiber strength in the axial direction is much greater than the horizontal, so for the fiber epoxy tube with circumferential winding, if not designed properly, the internal radial stress can easily exceed the fiber radial strength and resulting in layering and the failure of the entire rotor.

Fiber resin composite is a mixing of fiber and resin. Its mechanical properties are mainly provided by high-strength fibers. Resin combines the fibers and plays the role of the medium. The distribution of circumferential and radial stress depends on the value of μ of the composite material. The elastic modulus of a composite material is determined by the elastic modulus and composition ratio of its components.

$$E_\theta = E_f V_f + E_m V_m \tag{5.18}$$

$$E_r = E_f E_m / (V_m E_f + V_f E_m) \tag{5.19}$$

E_f—fiber modulus;

E_m—the amount of resin;

V_f—percentage of fiber volume;

V_m—the volume percentage of the resin.

In order to make the circumferential and radial stress of the composite rotor in the safety range so that the composite rotor will not fail at high-speed rotation, it is necessary to adjust the ratio of fiber and resin to obtain a reasonable value of μ to meet the conditions.

For components made of composite materials, there are a variety of safety evaluation criteria, including radial strength ratio R_r, circumferential intensity ratio R_θ, and Tsai-Wu failure criteria which is based on strength ratio R_T. FPF (First Ply Fail) and LPF (Last Ply Fail) also cannot be neglected. When FPF appears in the pulsed alternator composite rotor, the stiffness of the entire rotor changes thus brings imbalance and great vibration, which affects the rotor's normal operation at high speed. This provided, in the pulsed alternator, the strength criterion adopted should be FPF.

For radial strength and circumferential strength, the maximum stress in the composite flywheel is computed by analysis and calculation. The maximum stress point is used to verify the strength of the flywheel. Three strength criteria are calculated.

$$\begin{cases} R_r = \sigma_{r\max}/X \\ R_\theta = \sigma_{\theta\max}/Y \\ R_T = \sigma_{\theta\max}/Y \end{cases} \tag{5.20}$$

The safety factor is S_f:

$$S_f = \frac{1}{\max(R_r, R_\theta, R_T)} \tag{5.21}$$

5.3.2.2 Force Analysis of Composite Bandages

In order to prevent excessive centrifugal force in high-speed rotation, so the winding will not separate from the rotor yoke, the outer part of the winding is coated with carbon fiber of high strength density. Aiming to make the full of the advantages of carbon fiber lateral strength, taking the strength and cost of the material into consideration, the fiber is a monolayer ring winding. The circumferential carbon fiber bandage is equivalent to a cylindrical wall. The volume of the element is as shown in Fig. 5.10.

Ignoring axial stress, given that the force inside the bandage is flat stress, the strength matrix Q should be,

$$Q = S^{-1} = \begin{bmatrix} S_{rr} \; S_{r\theta} \\ S_{\theta r} \; S_{\theta\theta} \end{bmatrix}^{-1} = \begin{bmatrix} Q_{rr} \; Q_{r\theta} \\ Q_{\theta r} \; Q_{\theta\theta} \end{bmatrix} = \begin{bmatrix} \frac{E_\theta E_r}{E_r - E_\theta \nu_{\theta r}^2} & \frac{E_\theta E_r \nu_{\theta r}}{E_r - E_\theta \nu_{\theta r}^2} \\ \frac{E_\theta E_r \nu_{\theta r}}{E_r - E_\theta \nu_{\theta r}^2} & \frac{E_r^2}{E_r - E_\theta \nu_{\theta r}^2} \end{bmatrix} \tag{5.22}$$

$S_{\theta\theta} = 1/E_\theta$; $S_{\theta r} = S_{r\theta} = -\nu_{r\theta}/E_\theta$; $S_{rr} = 1/E_r$;

$\nu_{r\theta}$—primary Poisson's ratio;

$\nu_{\theta r}$—second Poisson's ratio, $\nu_{\theta r} = -\nu_{\theta r}\, E_r/E_\theta$.

Considering the residual stresses of temperature drop and chemical shrinkage, the radial stress is,

$$\sigma_r = -\rho\omega^2\varphi_1\bar{r}^2 + C_1\bar{r}^{k-1} + C_2\bar{r}^{-k-1} + \varphi_{T1}\Delta T_e \tag{5.23}$$

$\bar{r}$—the ratio of radius r to inner diameter r_i;

$\bar{r}_0$—the ratio of outer diameter r_o to inner diameter r_i;

ΔT_e—the curing temperature drop, equal to the room temperature T_0 minus curing temperature T_c;

k—$k = \sqrt{E_\theta/E_r}$, hardness ratio;

$\varphi_1 = \frac{3Q_{rr}+Q_{\theta r}}{(9-k^2)Q_{rr}}$; $\varphi_{T1} = \frac{\alpha_r-\alpha_\theta}{S_{\theta\theta}-S_{rr}}$;

$C_1 = \left(\frac{1-\bar{r}_0^{-k-1}}{\bar{r}_0^{-k-1}-\bar{r}_0^{k-1}}\right)\cdot\varphi_{T1}\cdot\Delta T_e$; $C_2 = -\varphi_{T1}\cdot\Delta T_e - C_1$.

The radius at the maximum radial stress $\bar{r}_{\sigma_r\_\max}$ can be obtained by taking the derivative to $\bar{r}$ from Formula (5.23).

$$\bar{r}_{\sigma_r\_\max} = \left(\frac{\bar{r}_0^{k-1}-1}{1-\bar{r}_0^{k-1}}\cdot\frac{k+1}{k-1}\right)^{\frac{1}{2k}} \tag{5.24}$$

As can be seen by Formula (5.24), the radius at which the radial maximum stress is produced and the maximum radial stress in the flywheel are related only to the material properties of the fiber and the ratio of the inner and outer diameter, and are independent to the actual size of the flywheel, so the scale of the flywheel in the actual engineering prototype can be reduced for calculation and experiments. The cost of design can be significantly reduced.

The bandage is directly wound on the excitation winding. The inner diameter of the bandage in the prototype is 150 mm, the thickness is 5 mm. In order to prevent the insulation of the winding at high temperature curing, while reducing the residual stress inside the bandage, the low temperature curing system is adopted, in which the curing temperature is 80 °C, room temperature is 20 °C, so the curing temperature is reduced to −60 °C. The lateral strength of fiber is much less than its longitudinal strength, so the strength of the fiber epoxy resin bandage in 90° is mainly concentrated in radial strength. The bandage is made of high-modulus high-strength T800 carbon fiber. The relationship between radial residual stress in the bandage after curing and the radial stress and the radius of the bandage within the bandage as shown in Fig. 5.11. In this case the bandage is rotated at high speed of 10,000 rpm.

As can be seen from Fig. 5.11, the internal residual stress at the time of curing has a maximum value at a certain point in the middle. However, due to the use of low temperature curing, the maximum value of residual radial stress is much lower than the radial strength of the material. The radial stress at the outer diameter is the largest

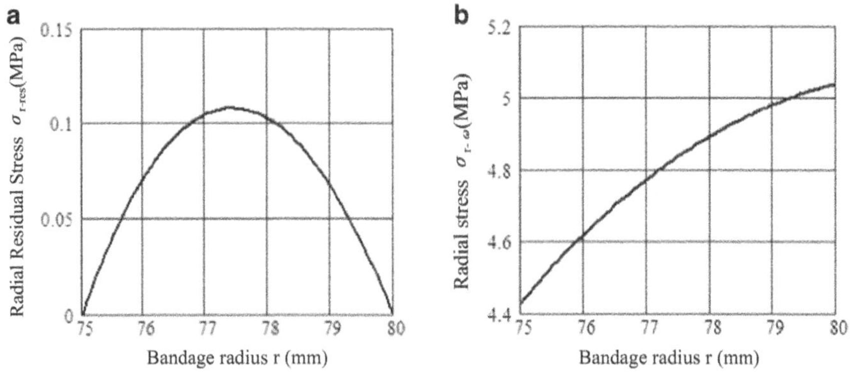

Fig. 5.11 During curing (**a**) and high-speed rotation (**b**) the curve of the change of the bandage radial stress with the radius

at the rotation of 10,000 rpm, which is 5.04 MPa, far less than the lateral strength of carbon fiber. There will be no stratification break, hence ensuring the safety of the rotor during high-speed rotation.

5.3.2.3 Force Analysis of Rotor Yokes

In the air core CPA, the fiber epoxy composite materials used are concentrated in fiber glass fiber and carbon fiber, and the performance parameters are shown in Table 5.2.

In previous air core CPA studies, air core rotors were made of a single composite material, aka glass or carbon fiber. The layering resulted from excessive flywheel radial stress impeded the normal operation of the entire rotor. In order to prevent this from happening, the flywheel uses a multi-wheel edge combined structure to replace a single rim. In addition, to prevent the failure of rotor layering at high speed and improve the energy storage density of air core rotors, the centrifugal stress during high-speed rotation is offset by the prestress. The multi-wheel edges are combined by interference fit, mainly by low temperature cooling of the internal ring or by room temperature pressure. In the design of CPA rotors for small-caliber electromagnetic guns, CEM scholars used the same carbon fiber to process several rings of different thickness, and cooled the internal ring to assemble the outer ring and realize the interference fit. The price of high-strength carbon fiber is dozens of times the price of glass fiber. For the sake of cost and material utilization, the fourth method can be used to design CPA's air core rotor, that is, to achieve the optimization of energy storage density with interference fit of different composite rings, as shown in Fig. 5.12.

For a single composite ring, it is believed that the internal residual stress has been reduced by reasonable winding and curing process. When designing the flywheel of multi-wheel edge structure, the interference magnitude in installing should be taken into consideration. If the interference magnitude is too large, the contact pressure between the adjacent interference surface of the flanges will be overwhelming,

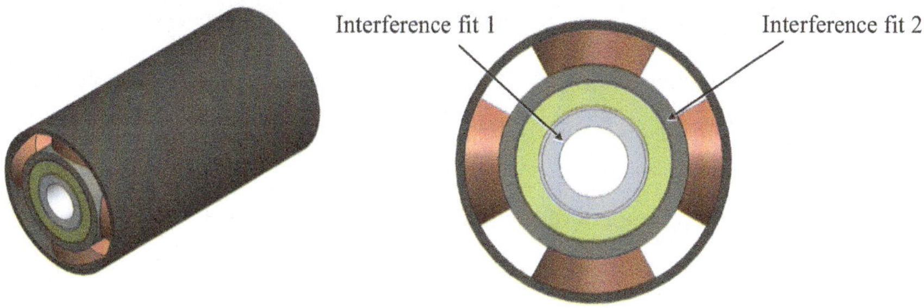

Fig. 5.12 3D and radial view of multi-wheel composite rotors

which will destroy the composite ring. If the interference magnitude is too small, the centrifugal force will not be fully offset, resulting in uneven distribution of radial stress inside the rim, then the purpose of preventing layering failure and improving the energy storage density of the rotor cannot be achieved.

With the improved plane strain FEM, assuming that there is a constant strain ε_0 in the axial direction, the stress distribution and energy density in the rotors of multi-wheel strains are calculated by the plane stress method, and the three-dimensional compliance matrix is as follows.

$$S = \begin{pmatrix} \frac{1}{E_\theta} & -\frac{\nu_{\theta r}}{E_r} & -\frac{\nu_{\theta z}}{E_z} \\ -\frac{\nu_{r\theta}}{E_\theta} & \frac{1}{E_r} & -\frac{\nu_{rz}}{E_z} \\ -\frac{\nu_{z\theta}}{E_\theta} & -\frac{\nu_{zr}}{E_r} & \frac{1}{E_z} \end{pmatrix} \tag{5.25}$$

After solving the plane stress equation, the radial displacement u within the ring i can be calculated as,

$$u^{(i)} = C_1^i \cdot b_i \cdot \chi^k + C_2^i \cdot b_i \cdot \chi^{-k} + \zeta_1 \cdot \varepsilon_0 \cdot \chi + \zeta_2 \cdot \Delta T \cdot b_i \cdot \chi + \zeta_3 \cdot \omega^2 \cdot \chi^3 \cdot b_i^3 \tag{5.26}$$

Q is the hardness matrix and the inverse matrix of the compliance matrix S; $\boldsymbol{\beta} = Q \cdot \boldsymbol{\alpha}$; $\zeta_1 = (Q_{13} - Q_{23})/(Q_{22} - Q_{11})$; $\zeta_2 = (\beta_2 - \beta_1)/(Q_{22} - Q_{11})$; $\zeta_3 = -\rho/\left((9 - k^2)Q_{22}\right)$.

Assuming that the internal and outer diameters of the multi-wheel rims are stress-free and in a free state, it can be calculated according to the following boundary conditions.

$$\begin{cases} \sigma_r^{(1)}(r = a_1) = 0 \\ \sigma_r^{(n)}(r = b_n) = 0 \\ \sigma_r^{(i)}(r = b_i) = \sigma_r^{(i+1)}(r = a_{i+1}) \\ \sum_1^n \int_{c_i}^1 b_i^2 \sigma_z^{(i)} \chi^{(i)} d\chi^{(i)} = 0 \\ \delta^{(j)} = u_{r_i}^{(j+1)} - u_{r_o}^{(j)} \end{cases} \tag{5.27}$$

n—the number of layers that consists the flywheel;

a_i, b_i—the inner and outer diameter of layer i of the composite material respectively;

c_i—the ratio of the inner diameter to the outer diameter of layer i of the composite material;

$u_{ri}{}^{(j+1)}$—the radial displacement of the inner diameter of the $j+1$ layer;

$u_{ro}{}^{(j)}$—the radial displacement of the inner diameter of the j layer;

$\delta(j)$—the magnitude of interference of the layer j and $j+1$.

According to Formula (5.27), the parameters $C_1{}^{(i)}$ and $C_2{}^{(i)}$ and axial strain ε_0 can be obtained, and the radial stress $\sigma_r{}^{(i)}(\chi)$ can be obtained in the lower formula:

$$\begin{aligned}\sigma_r^{(i)}(\chi) = {} & C_1^{(i)}(Q_{12} + kQ_{22})\chi^{k-1} + C_2^{(i)}(Q_{12} + kQ_{22})\chi^{-k-1} \\ & + \varepsilon_0[\zeta_1(Q_{12} + Q_{22}) + Q_{23}] + \Delta T[\zeta_2(Q_{12} + Q_{22}) - \beta_2] \\ & + \zeta_3\omega^2(Q_{12} + 3Q_{22})\chi^2 b_i^2\end{aligned} \tag{5.28}$$

Taking the flywheel made from two layers of composite rings as an example, the stress analysis of the three structures is carried out as follows: Structure I is made of two composite tubes of equal thickness in an over-assembly way, each made of single T800 carbon fiber through firing. Structure II and Structure III are made of two composite tubes with the equal thickness to structure I in an over-assembly way. Each composite tube is entangled with different fibers. The outer one is made of T800 which boasts larger modulus and higher strength. The interior is made of T300 whose modulus is smaller and strength is lower. The four flywheel structures have the same size, with a 500 mm internal diameter 500 mm, an inner ring thickness of 100 mm, an outer ring thickness of 100 mm, a 1 mm magnitude of interference between rings, and a curing temperature drop of −60 °C. The comparison of the performances of the three structures are shown in Tables 5.3.

Since the circumferential strength of the fiber is much greater than the radial strength, only the core radial stress is required to be verified in the 90° toroidal twisting flywheel. The safety factor F_r is,

Table 5.3 Comparison of flywheel performance with different composite material structures under the same sizes

Performance metrics	Structure I	Structure II	Structure III
Material of inner and outer rings	T800 + T800	T300 + T800	GF + T800
Radial maximum strength (MPa) (10,000 rpm)	47	37	−35
Rotor mass (kg)	1206	1206	1275
Maximum rotating speed (rpm)	16,500	23,000	16,500
Maximum energy storage (MJ)	666	1295	698
Maximum energy storage density (kJ/kg)	552	1073	547

$$F_r = \begin{cases} \sigma_r/Y, \sigma_r > 0 \\ \sigma_r/Y', \sigma_r < 0 \end{cases} \tag{5.29}$$

The radial stress and safety coefficient distributions of the three different structures at designed speeds of 10,000 rpm are shown in Fig. 5.13.

From Fig. 5.13, it can be figured that in the composite flywheel of the same material, such as structure I, internal radial stress equals to the tensile stress, so it is more likely for the radial strength failure to happen and result in layering. This is because the radial tensile strength of the fiber is less than radial compression strength. Composite flywheels with different materials provide better performance. The internal radial stress of structure II is less than structure I, with maximum limit speed and energy storage density. Although its maximum limit speed is less than structure II, the internal radial stress of structure III is equal to the pressure stress, which avoids the risk of radial layering at high speed, therefore reducing the requirement to the magnitude of interference during assembly, thus lowering the installation difficulty. In addition, the material cost of the structure is lower than that of structure I and structure II. This is another comprehensive advantage.

5.4 Study on the Mechanical Characteristics of Pulsed Alternators

Rotor dynamics is the study of all rotating mechanical rotors and their components and structure-related mechanical characteristics, no matter how precise the dynamic balance of the rotor is, there is always a certain deviation between the rotor's center of mass and gyration, which leads to the periodic centrifugal interference in the rotor. When the rotor's speed is close to the critical speed of the rotor, the rotor will have a violent bending vibration, causing the whole unit to vibrate and, in severe cases, the rotor could possibly break. Therefore, it is necessary to make a study of the mechanical characteristics of the rotor system, accurately analyze the critical speed of the rotor system, so that the working speed of the rotor will be far from the critical speed under the bending mode, so as to ensure the safety and stable operation of the rotor.

5.4.1 Status of High-Speed Rotor Dynamics Research

Rotor dynamics is a branch of solid mechanics, which is to study on the critical speed of the rotor system of high-speed rotating machinery in working state, the unstable state of the rotor through critical rotational speed, the dynamic response, the dynamic balance, the rotor stability and so on. The engineering and science communities have been concerned about rotor vibration for more than 200 years. In 1869, Rankine from

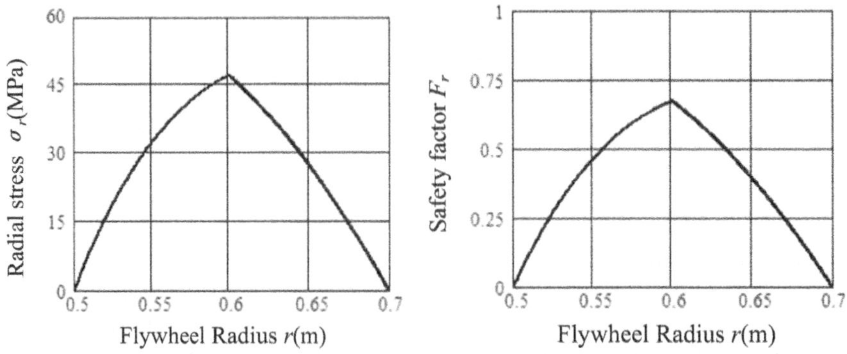

a) Radial stress and safety factor of flywheel I of T800+T800 hybrid structure

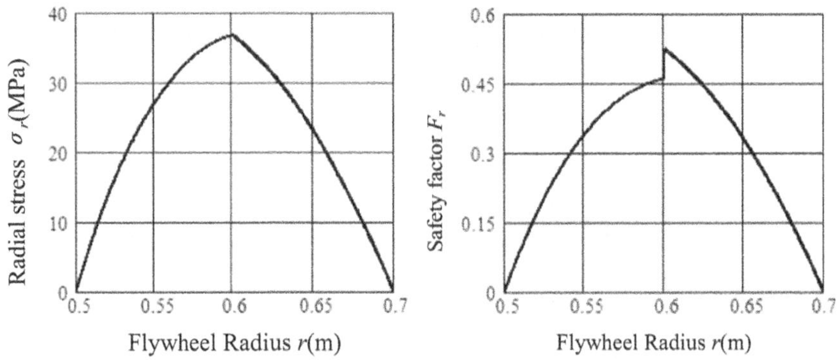

b) Radial stress and safety factor of flywheel I of T300+T800 hybrid structure

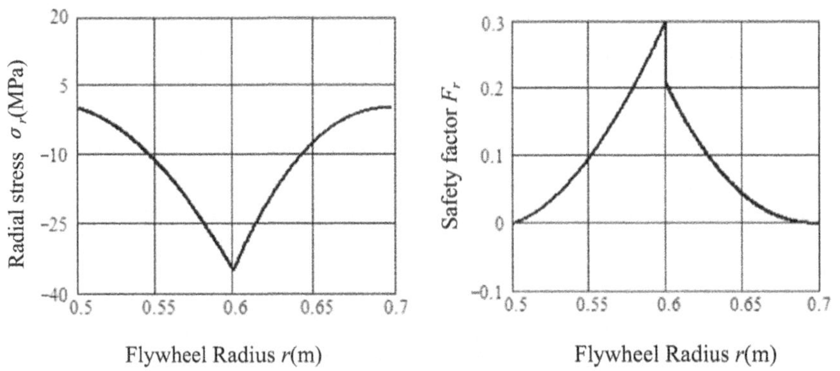

c) Radial stress and safety factor of flywheel I of GF+T800 hybrid structure

Fig. 5.13 Stress assessment of three composite flywheels of hybrid structures

England published his paper *On Rotating Centrifugal Force*, which kept the rotor confined to first critical street speed for a long time without progress; in 1889, French scientist Laval initiated the experiment on flexible shafts and piloted the study in this field. The simplest rotor model proposed by Foppl in 1895, which was a massless rotor with two rigid ends and a disc in the middle; in 1919, JeffCott first explained that the rotor would produce automatic fixation in supercritical operation and stabilize the process. Last but not least, in a landmark international conference on rotor dynamics held in Denmark in 1974, it was pointed out that the research on rotor dynamics mainly includes five aspects: critical speed, the unstable state of the rotor through the critical rotational speed, dynamic response, dynamic balance and rotor stability. These are all what the traditional rotor dynamics concerns about.

With the continuous development of science and technology and the continuous improvement of the rotor speed of rotating machinery, it is more and more important to study on the dynamics of the rotor, such as the improvement of high-speed motor speed and its vibration and stability problems of high-speed rotor system. When discussing and studying the dynamic performance of rotor dynamics, with the continuous improvement of rotor speed, it is necessary to carry out the computational analysis of rotor dynamics on rotating machinery in the design of rotor and rotor system. In this way, the computational analysis of rotor dynamics is the primary guide of engineering practice. In the course of the development of rotor dynamics, there are many methods for computational analysis, but they can be mainly divided into two main categories: the transfer matrix method and the finite element method. The transfer matrix method has played a leading role in the field of rotor dynamics, the advantage of which is that the matrix order does not increase with the rising freedom of the rotor system. This is quite simple and efficient. If combined with other methods, many dynamic problems of complex rotor systems can be solved. It's also suitable for chain rotor systems with multiple rotors connected with each other. However, because the transfer matrix is to simplify the relevant quality of the rotor system structure and change it into an extremely simple concentrated-mass and massless beam model, the disadvantage is that the integrity and analysis of the model cannot be guaranteed, and sometimes the solution could be missing. When calculating rotor dynamics problems with the finite element method, the deficiency of the transfer matrix method can be fixed and the completeness, accuracy and efficiency can be guaranteed. The rotor dynamics equation form the finite element method is fairly clear, and the calculating results are more accurate than the transfer matrix method when solving complex problems. The finite element method can analyze the Bernoulli beam and Timoshenko beam by linearizing the rotor system and seeking the load. Although the calculation of finite element method is time consuming, with the development of modern science and technology and the popularization of computers, it provides a good hardware support technology for finite element method and makes it widely available at present.

5.4.2 Critical Speed and Modal Analysis

Due to errors in manufacturing, the centroid of each micro-segment of the rotor generally deviates slightly from the rotation axis. As the rotor rotates, the centrifugal force caused by the above deviation makes the rotor vibrate laterally. This vibration appears to be unusually strong at some rotation speeds, which are called critical speeds. In order to ensure that the machine does not resonate within the operating speed range, the critical speed should be appropriately deviated from the operating speed by, for example, more than 10%. Critical speed is related to factors such as the elasticity and mass distribution of the rotor. For discrete rotation systems with limited centralized mass, the number of critical speeds is equal to the number of concentrated masses. While for elastic rotation systems with continuous distribution of mass, there are infinite critical speeds. The most common numerical method for calculating the critical speed of large rotor support system is the transfer matrix. The main point is: first, to divide the rotor into several segments, the relationship between the four cross-section parameters (deflection, angular deflection, bending moment, and shear) at the left and right end of each segment can be described by the transfer matrix of that segment. In this way, the total transfer matrix between the cross-section parameters of the left and right ends of the system is obtained. Then by the boundary conditions and the non-zero solution conditions during inherent vibration, the trial-and-error method is adopted to obtain the critical speed of each order, and also the corresponding vibration type. General rotors go through the critical speed at a variable speed, so the state in which the rotors go through the critical speed is not stable It is different from the stable state of rotation fixed at critical speed in two aspects: first, the maximum value of amplitude is smaller than the one in stable state, and the faster it rotates, the greater the value of amplitude is; the second is that the great value of amplitude does not occur at critical speed as it is in a stable state. In the state of instability, a variable frequency interference force influences the rotors and makes the analysis difficult. To solve such problems, it is necessary to use numerical calculation or asymptomatic method in nonlinear vibration theory. Sometimes series expansion methods may also come into use.

Modals refer to the characteristics inherent in the frequency and vibration pattern of the structure. The process of extracting the modal characteristic parameters of each mode by computational analysis is called modal analysis. Modal analysis is the starting point or basis for dynamic analysis problems such as harmonic response analysis and transient analysis. There are two kinds of modal analysis methods: one is the finite element method, and the other is experimental method. Modal analysis is a problem in the aspect of the characteristic value and characteristic vector of the rotor system. Among them, the characteristic vector refers to the modal vibration pattern of the structure vibration of the rotor system, and the characteristic value refers to the frequency corresponding to the modal vibration pattern of the structure vibration of the rotor system. In practical conditions, sometimes the rotation speed is avoided to take place near these frequencies to prevent resonance, while sometimes, in very special circumstances, vibration is required to be enhanced. It depends on specific conditions. In summary, according to the rotor natural frequency and other modal parameters from the modal analysis, a determination criterion can be given in the design of the rotor system.

Chapter 6
Controlling Technology for Pulsed Alternators

6.1 Overview of Pulsed Alternator's Measurement and Controlling System

The measurement and control system is the auxiliary device of the pulsed alternator power supply system. Its main task is to complete the monitoring and protection of the system reliably. This includes monitoring system and control system. The operation process of the power supply system of the pulsed alternator is very short. The discharging only lasts for a few milliseconds. Therefore, in this process, the operator doesn't have enough time to participate in the real-time observation. The role of the monitoring system is to record the relevant data of the discharge process for subsequent analysis. The control system of pulsed alternator is responsible for the field control and protection of the whole system, and the difficulty is to guarantee real-time and reliability.

The specific function of the measurement and control system mainly includes the control of overall work process, fault handling and protection and pulsed power output waveform regulation. The first step is to control the operation process of the air core pulsed alternator. According to the pre-experimental design, the motor excitation and control system will control the motor's working process, so that it can complete the entire motor work according to the predetermined timing and logical relation. Secondly, the system will also implement all necessary protection of the motor to deal with various conditions. As long as there appears excessive speed, overcurrent and overheat of the motor in its operation, the system must provide reliable protection to the motor, in order to avoid motor damage or casualties. Finally, the purpose of the air core pulsed alternator system is to provide the pulsed current to meet the technical requirements for various types of loads. Because the load has specific technical requirements for the output current waveform of the motor, in addition to the operating sequential control and protection control which are same as conventional systems, it's also necessary for the control system to adjust the output of the motor to match with the specific load requirements.

S. Wu and S. Cui, *Pulsed Alternators Technologies and Application*,
https://doi.org/10.1007/978-981-33-4224-8_6

According to the characteristics of the compensated pulsed alternator system and the design experience of the actual project, the design objectives of the excitation control system are formulated. In order to compare and analyze different schemes, the following requirements are put forward for excitation and control systems:

(1) High power grade

Ferromagnetic materials are used in conventional electro-excitation synchronous generator. Because of the high magnetic conductivity of ferromagnetic materials, only a small excitation power can produce the required excitation magnetic field. For the air core compensates the pulsed alternator, non-ferromagnetic materials are selected, and the motor magnetic conductivity is close to the magnetic conductivity in the vacuum. Therefore, in order to establish an excitation magnetic field with required strength, it is necessary to provide a large excitation power, which puts forward higher requirements for the power grade of the measurement and control system to prepare favorable insulation and protection.

In order to obtain the largest possible output current, the excitation system must provide excitation power to meet certain requirements and maximize the advantages of the empty core pulse generator. After adjusting the excitation and control system, it is necessary to obtain the required current waveform. In addition, the high frequency discharge of the air core pulsed alternator also brings about the requirements for the excitation and control system.

(2) High reliability

Reliability includes the reliability under normal operation or fault. The former refers to a streamlined scale of the system, i.e. the amount and types of design components need to be reduced to prevent failure. The latter refers to the quick response of the system to the emergence of excessive speed, excessive current and other undesired states by starting the protection program instantly to avoid accidents. The system itself also requires reliability, while reducing the complexity of control by minimizing the number and scale of components required for the system for optimization. After meeting the above requirements, it is also necessary to take into account the feasibility of the principle of the system program. The feasibility of the principle can be verified by theoretical analysis and software simulation, while the latter two require detailed market research in system design. It's preferable to carry out design and research at the same time. Only by considering the feasibility requirements can the design become a practical experimental device.

(3) Keep the weight and volume as small as possible

Portability is a concept for the pulsed alternator power supply system. One of the advantages of compensating pulsed alternator is to meet the power density and energy density under the premise of the unit advantage, so in this design, the size and weight of components need to be taken into account. As a new type of pulsed power supply, the air core pulsed alternator has a higher energy density than other forms of pulsed power supply, such as capacitors, so it is also required to keep the weight and volume of the excitation control system of the motor as small as possible. In this way, the advantage of the alternator will not be offset by the volume and weight of the auxiliary device. In the whole scheme design and the process of device selection, the volume and weight of the whole system need to be considered.

6.2 Excitation Process

Excitation control system is an important part of the power supply system of pulsed alternator, its core task is to provide excitation power and control the excitation process of the motor. For the optimal performance of the generator, the task of the excitation control system is to ensure reliable motor excitation current. Because the ferromagnetic material used to manufacture conventional motors has a high magnetic conductivity, it is only necessary to provide a relatively small excitation power to fully establish the required magnetic field strength in the motor air gap. But for the air core pulsed alternator, non-magnetic materials were used in the manufacture. In order to form a higher magnetic field strength in the motor air gap, it requires a very large excitation power. In terms of the energy source of the excitation magnetic field, it can be divided into high-voltage capacitor energy supply, prime mover energy supply and rotor's own mechanical energy supply.

6.2.1 High-Voltage Capacitors Provide Excitation Magnetic Field Energy

6.2.1.1 Single Pulse Excitation

Single pulse excitation scheme is one of the simplest excitation schemes in structure, as shown in Fig. 6.1. A capacitor is paralleled at both ends of the generator's excitation windings to achieve motor excitation by controlling the capacitor's discharge of the motor excitation winding. The series of motor armature and discharging thyristor

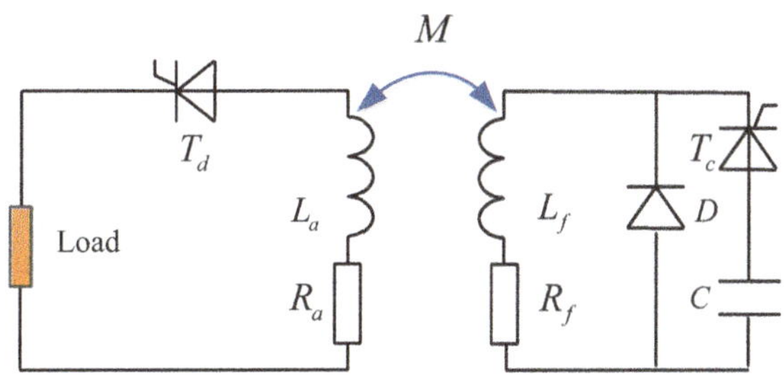

Fig. 6.1 Single pulse excitation scheme

control the driving load, and the inversely paralleled diodes of the motor excitation armature work for after flow.

When the system is working, the capacitor is first charged to a certain value by the capacitor charging power supply. The air core pulsed alternator was driven to the rated speed by the controlling prime mover, and then the system detects the rotation speed and the position of the motor rotor in real time. When the motor rotates to the appropriate position, the system will trigger the excitation discharge switch, the motor excitation armature winding will be charged by the capacitor, an internal magnetic field will be built in the motor. When the excitation current reaches the predetermined value, the main circuit will be discharged through the armature winding discharging switch.

The entire excitation process is divided into two phases:

1. The high voltage capacitor discharges to the excitation armature winding through the thyristor.

As shown in Fig. 6.2, a diagram of the charging of the high-voltage capacitor to the excitation winding, there is

$$R_f C\frac{du_C}{dt} + L_f C\frac{d^2u_C}{dt^2} + u_C = 0 \tag{6.1}$$

The initial condition is

$$\begin{cases} u_C(0_+) = u_C(0_-) = U_o \\ \frac{du_C}{dt}|_{t=0_+} = \frac{1}{C}i(0_+) = \frac{1}{C}i(0_+) = 0 \end{cases} \tag{6.2}$$

Taking $R_f < 2\sqrt{\frac{L_f}{C}}$,

$$u_C = \frac{U_o\omega_o}{\omega}e^{-\delta t}\sin(\omega t + \beta) \tag{6.3}$$

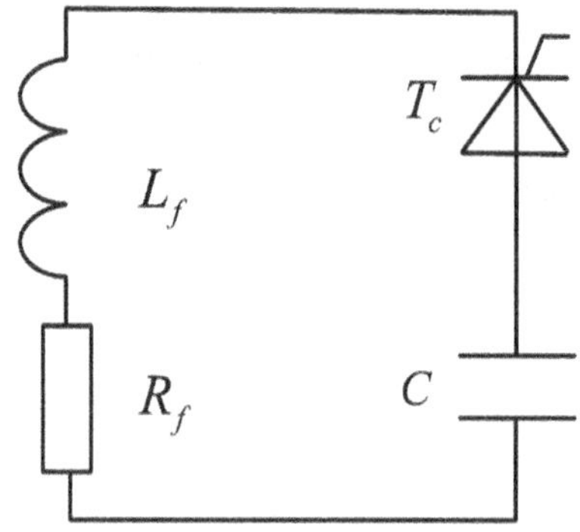

Fig. 6.2 Diagram of high-voltage capacitor charging excitation armature windings

$\omega = \sqrt{\frac{1}{L_f C} - (\frac{R_f}{2L_f})^2}$, $\delta = \frac{R_f}{2L_f}$, $\omega_o = \frac{1}{\sqrt{L_f C}}$, $\beta = \arccos \frac{\delta}{\omega_o}$, U_o is the initial voltage of the capacitor.

When the capacitor voltage is zero, the capacitor stops charging the excitation winding, which makes the excitation armature winding enter the after flow. This is as shown in the following (Figs. 6.3 and 6.4).

In this process,

$$\omega t + \beta = \pi \tag{6.4}$$

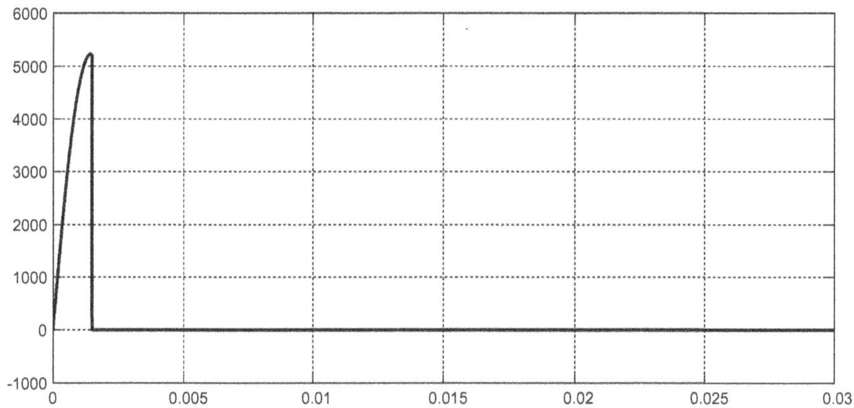

Fig. 6.3 Capacitor discharging current

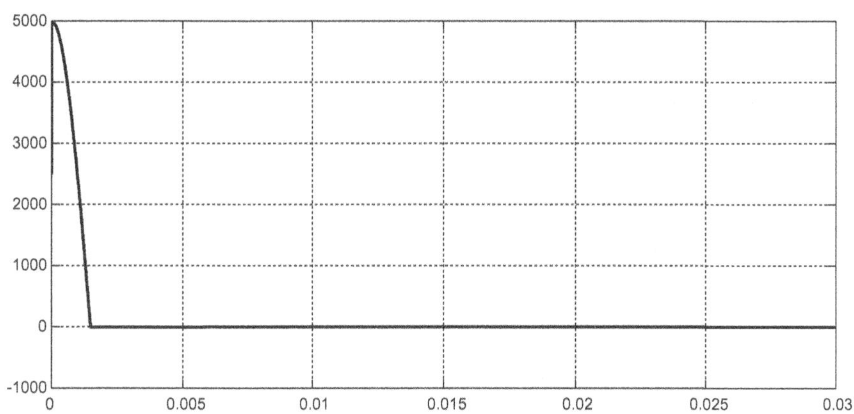

Fig. 6.4 The voltage at the two ends of the capacitor

$$t_{Ch\arg e} = \frac{\pi - \beta}{\omega} = \frac{\pi - \arccos\frac{R_f}{2}\sqrt{\frac{C}{L_f}}}{\sqrt{\frac{1}{L_f C} - (\frac{R_f}{2L_f})^2}} \tag{6.5}$$

$L_f = 843.6\,\mu H$, $R_f = 48.3\,\text{m}\Omega$, $\text{C} = 1\text{mF}$, $t_{Ch\arg e} \approx 1.47$ ms can be calculated.

$$i = C\frac{du_C}{dt} = \frac{U_o}{\omega L_f}e^{-\delta t}\sin(\omega t) \tag{6.6}$$

The capacitor discharging current is

$$i(t_{Ch\arg e}) = \frac{U_o}{\omega L_f}e^{-0.00147\delta}\sin(0.00147\omega) = 5218.93\text{ A} \tag{6.7}$$

After calculating the extreme value based on the current expression, when the current $t = \frac{\beta}{\omega} = 1.42$ ms, the current reaches the maximum value, which is

$$i_{\max} = \frac{U_o}{\omega L_f}e^{-\frac{\delta\beta}{\omega}}\sin\beta = 5224.31\text{ A} \tag{6.8}$$

After 1.47 ms, the excitation winding enters the phase of after flow, and the current at the beginning of this stage is 5218.93 A, which is less than the maximum current of the capacitor to discharge to the excitation armature winding.

1. Excitation armature windings and the fly-wheel diode compose a continuous current circuit which is ready to discharge (Fig. 6.5).

When the capacitor voltage is zero, the thyristor stops working, after which it enters the phase of after flow, ready to discharge the load.

$$L_f\frac{di_L}{dt} + R_f i_L = 0 \tag{6.9}$$

$$i_L(0_+) = i_L(0_-) = i_o = i(t_{Ch\arg e}) \tag{6.10}$$

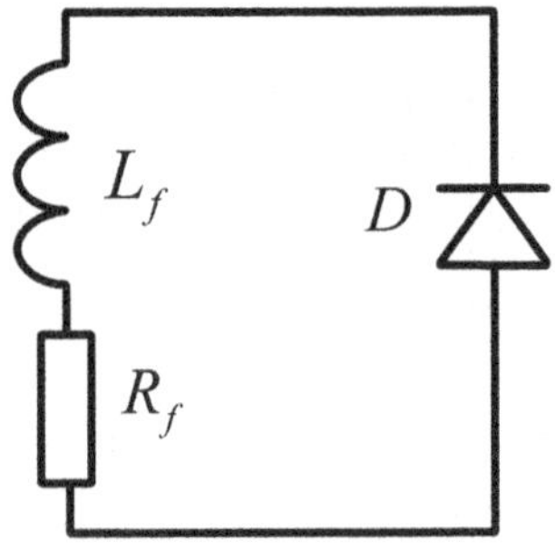

Fig. 6.5 Continuous circuit of the excitation winding

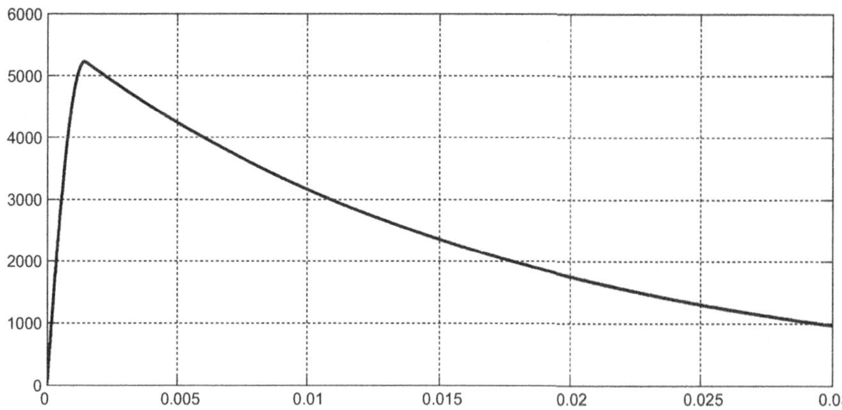

Fig. 6.6 Excitation winding current

$$i_L(t) = Ae^{-\frac{R_f}{L_f}t} \tag{6.11}$$

$$i_L(0_+) = Ae^o = A = i_o \tag{6.12}$$

$$i_L(t) = i_o e^{-\frac{R_f}{L_f}t} \tag{6.13}$$

The current in the excitation winding is attenuated exponentially from the initial value, whose rate depends on the inductance and resistance of the excitation armature winding. The greater the inductance is, the smaller the resistance will be, and the longer the transient process will be. The change of the current of the excitation winding is as shown in the Fig. 6.6.

The advantage of the excitation scheme is that the simple structure of the system topology, the small amount of the switching component, the accessible controlling method, the high reliability. All of these can provide continuous current pulse, which are beneficial in some particular applications.

6.2.1.2 Multi-Pulse Excitation

The most useful point of excitation current is to provide an excitation magnetic field when the armature winding discharge occurs, so that during non-discharge, the excitation current is useless, only to increase the system's loss. If a single pulse excitation scheme is adopted, in the non-discharge time, there is still current in the excitation winding, resulting in a larger overall loss of the system. Therefore, multi-pulse excitation scheme is favorable to produce pulsed excitation current in the required discharging time. In the non-discharge time, the current in the excitation winding is absent.

Figure 6.7 is a schematic that shows how high-voltage capacitors and excitation windings form oscillation circuits, and the current of the excitation circuit is as shown in Fig. 6.8.

In order to give the system greater control, switching components need to be added in the excitation windings, as shown in Fig. 6.9. The scheme retains the thyristor of the main circuit discharging switch, removes the fly-wheel diode at both ends of the motor excitation winding, and adds a diode component which is in reverse parallel with the original excitation circuit discharging switch.

The multi-pulse excitation mode mainly has the following steps:

(1) The control system issues instructions to charge the excitation capacitor voltage to the rated value by using the charging power supply;
(2) The start-up prime mover drives the air core pulsed alternator to the rated speed and tests the speed and position of the rotor.

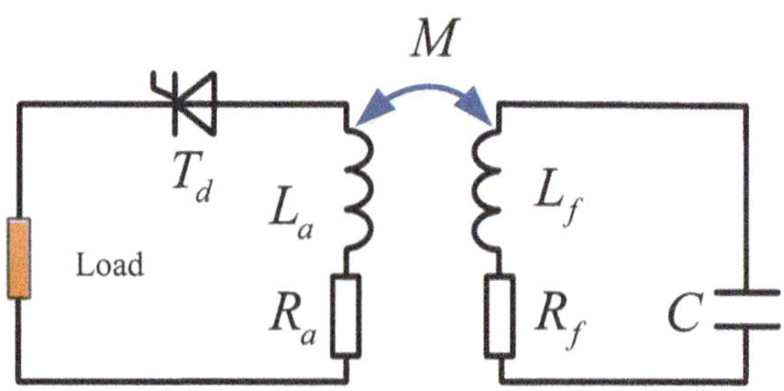

Fig. 6.7 High voltage capacitors and oscillating circuit

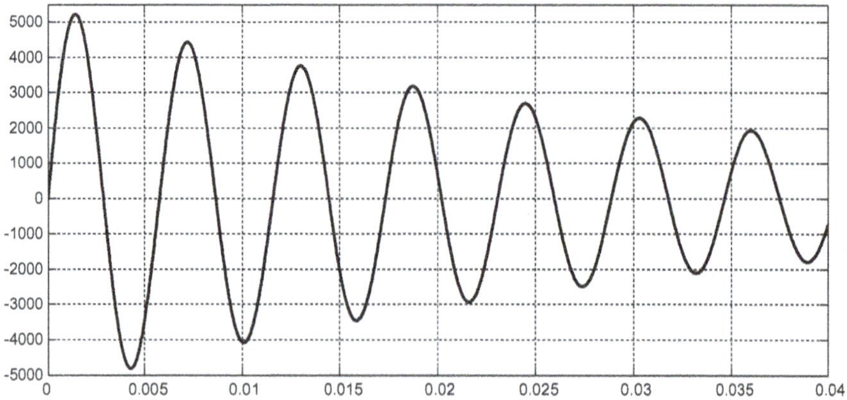

Fig. 6.8 Excitation circuit current

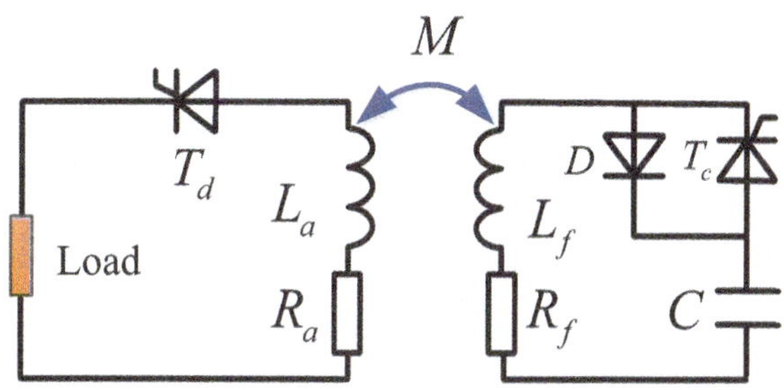

Fig. 6.9 Optimization of the circuit topology

(3) After the discharge condition is met, when the system detects that the motor has turned to a suitable angle, the discharge discharging thyristor switch will be triggered, and the capacitor will excite the motor excitation winding. In this way the discharging switch of the main circuit will be triggered in due course during the discharge of the capacitor to the excitation winding.
(4) The excitation winding will charge the capacitor in reverse by a diode in reverse to the wire after discharge, at which point the thyristor assembly is naturally shut down due to reverse voltage, and when the energy stored in the motor excitation winding is fully converted to the energy in the capacitor, the current of the motor excitation winding is zero.

The above process will be repeated until the excitation current decays to the point where the need for the discharge excitation cannot be met. Figure 6.10 is the multi-pulse excitation current waveform with more switches.

The physical essence of the multi-pulse excitation scheme is to utilize the inductor of the motor excitation winding and the oscillation of the excitation capacitor to achieve the repeated excitation of the motor, and the initial storage energy in the

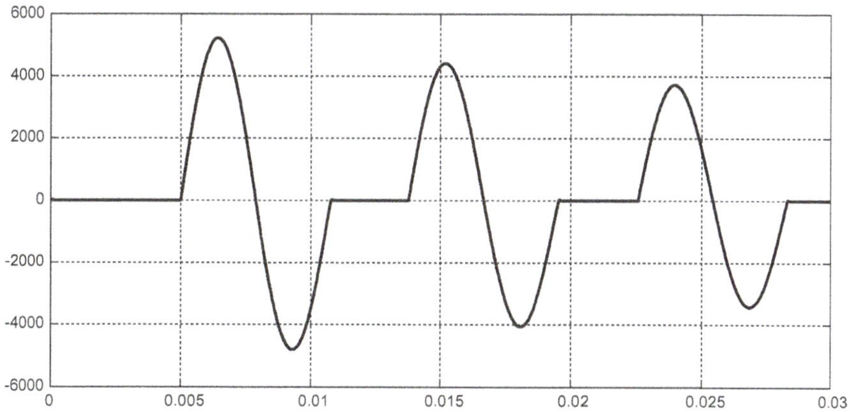

Fig. 6.10 Multi-pulse excitation current waveform with more switches

excitation capacitor is exchanged repeatedly between the excitation winding inductor and the capacitor of the motor. Parameters like the excitation current cycle and amplitude of the multi-pulse excitation scheme are determined by the excitation winding inductor, resistance and capacitor tolerance.

Multi-pulse excitation scheme can be used in some occasions with high requirements to the operating pulsed frequency. Once the motor's calorific value is too large or the motor speed drops too fast, it can be optimized through the controlling system regulation by reducing discharging frequency, so this scheme has a great flexibility. However, due to the increase of the complexity of the controlling system, the advantages of this scheme in terms of engineering cost and feasibility are no longer obvious. The improvement of the overall system performance of excitation capacitors has a great dependence.

6.2.2 The Rotor's Own Movement Provides Excitation Magnetic Field Energy

The excitation currents of the above two excitation schemes are provided by the external excitation capacitors, so it is called his excitation scheme. If improvement of the overall performance of the motor needs to be carried out, it's necessary to increase the excitation current of the motor, in this way it requires higher capacity of the additional excitation capacitor. Therefore, the volume of the whole system is increased, which means it cannot take advantages of the unit. The energy density and power density of the whole system will be reduced, and the flexibility of the system will also be limited. In order to obtain a higher motor excitation current without increasing the weight and volume of the system, self-excitation schemes can be used to reduce the importance of excitation capacitors.

The excitation process of the air core pulsed alternator is generally divided into three stages: first, the initialization (charging) of the hoist capacitor stage; second, the unstable stage of the rise of the excitation current index; third, the transition to a steady state process, which is achieved by the solid-state switching device. Specifically, first, the excitation capacitor needs to be charged to the voltage required by the system. Second, the capacitor discharges to the excitation winding and starts the positive feedback from the excitation process. The excitation current grows exponentially to establish the required air gap magnetic field strength. Finally, through the switching of the switch state, there will be a transition to the after flow. The armature winding will be the open circuit, and the excitation circuit will reflow the current with the continuous flow circuit.

6.2.2.1 Circuit Topology

In the power supply mode of the excitation winding, the SCC has set up a set of windings dedicated to self-excitation on the motor rotor, which is perpendicular to the axis of all the other windings at an electrical angle, in order to prevent coupling between the two sets of windings. The system uses a half-controlled rectifier bridge, and the excitation winding is continuous flowed by a continuous-flow diode bridge arm after the motor discharge process is completed (Figs. 6.11 and 6.12).

Compared with the SCC scheme, RGC adds a set of armatures, so that the two sets of windings on the motor rotor are completely separated, which brings about a dedicated demagnetization circuit for the motor. The residual energy of the motor will be consumed on the demagnetization resistance shown in the figure, thus reducing the heat of the motor excitation winding, which is conducive to the reliable work of the motor (Fig. 6.13).

There is only one set of armature windings on the rotor of CCEMG, which are used for excitation in the motor excitation link and discharging the load after the

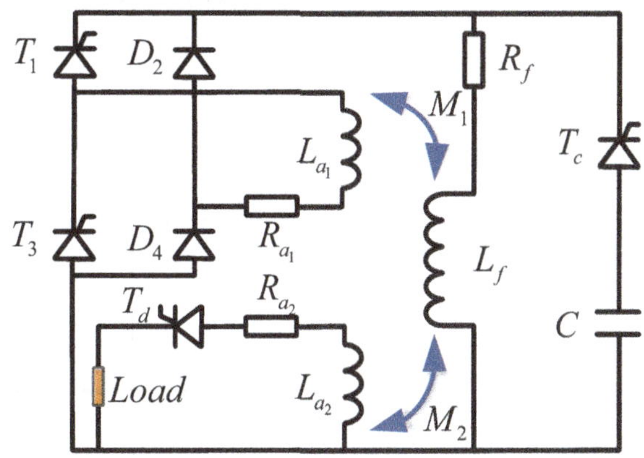

Fig. 6.11 Double armature winding's single-phased half-controlled rectifier

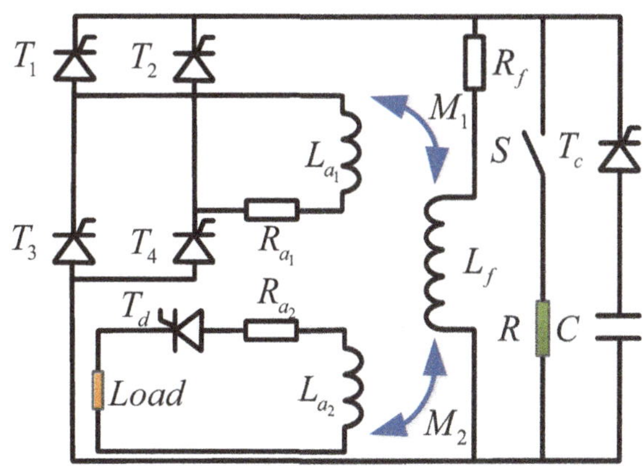

Fig. 6.12 Double armature winding single-phased full-control rectifier

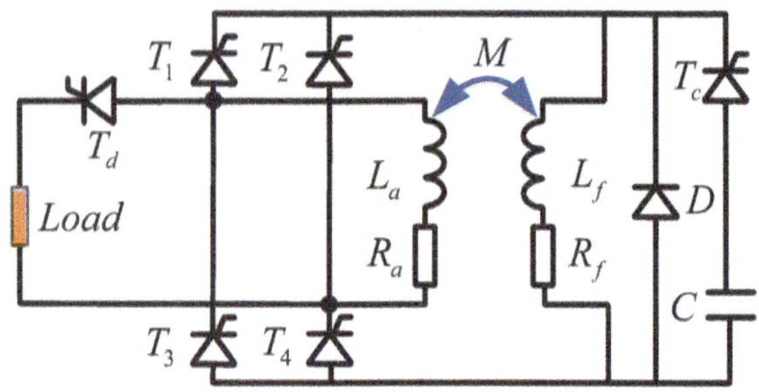

Fig. 6.13 Single armature single-phased full-controlled circuit

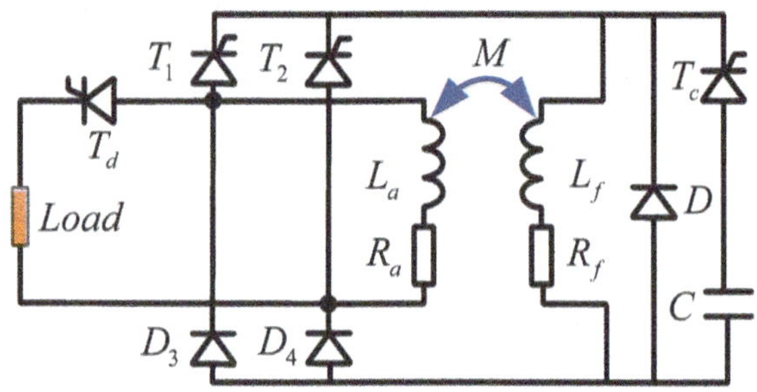

Fig. 6.14 Single armature single-phased half-controlled circuit

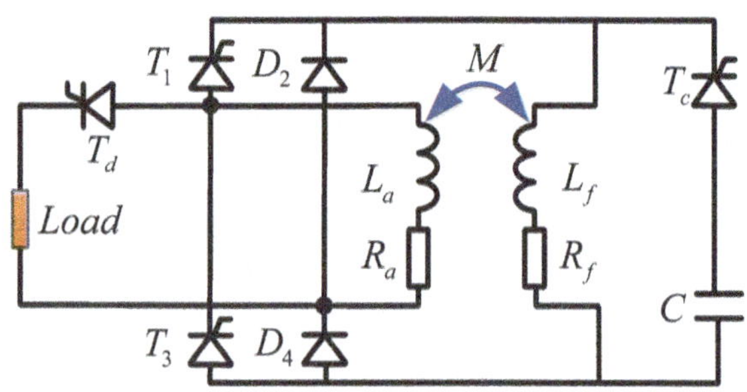

Fig. 6.15 Single armature single-phased half-controlled circuit

motor excitation. Simplification can be seen in the design. A full-controlled bridge is adopted, and a dedicated excitation flow-on diode is added (Figs. 6.14 and 6.15).

6.2.2.2 Mathematical Model

The definition of symbols in the formula are as follows. u_a is the armature winding voltage, i_a is the excitation winding voltage, i_a is the armature winding current, i_f is the excitation winding current, θ is the electrical angle between armature winding and excitation winding, R_a is the armature winding resistance, R_f is the excitation winding resistance, L_a is the armature winding inductance, L_f is the excitation

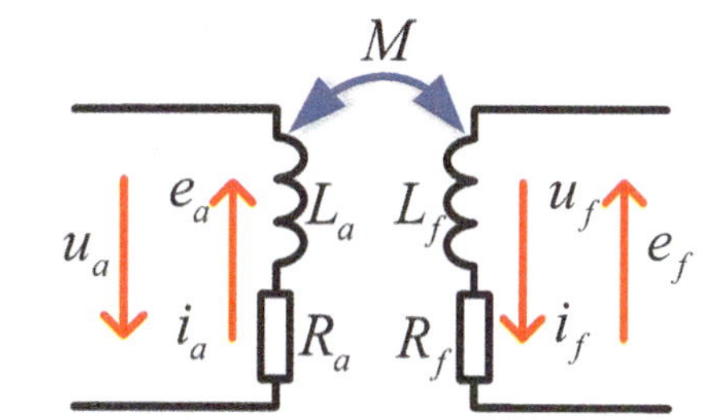

Fig. 6.16 Simplified circuit diagram

winding inductance, M is the largest mutual inductance between armature windings and excitation windings, p is the role pair number of the motor (Fig. 6.16).

According to the convention of the motors, positive current generates negative flux in armature windings, while bringing about positive flux in excitation windings.

The flux equation matrix of the motor is

$$\begin{bmatrix} \psi_a \\ \psi_f \end{bmatrix} = \begin{bmatrix} L_a & M_{af} \\ M_{fa} & L_f \end{bmatrix} \begin{bmatrix} -i_a \\ i_f \end{bmatrix} \tag{6.14}$$

The voltage equation of the motor is

$$\begin{bmatrix} u_a \\ u_f \end{bmatrix} = p \begin{bmatrix} \psi_a \\ \psi_f \end{bmatrix} + \begin{bmatrix} R_a & \\ & L_f \end{bmatrix} \begin{bmatrix} -i_a \\ i_f \end{bmatrix} \tag{6.15}$$

The total magnetic field energy of the motor is

$$\begin{aligned} W_m &= \frac{1}{2} I^T L I = \frac{1}{2} \begin{bmatrix} -i_a \\ i_f \end{bmatrix} \begin{bmatrix} L_a & M_{af} \\ M_{fa} & L_f \end{bmatrix} [-i_a \ i_f] \\ &= \frac{1}{2} L_a i_a^2 + \frac{1}{2} L_f i_f^2 - M i_f i_a \cos\theta \end{aligned} \tag{6.16}$$

According to the "theorem of virtual displacement", when the current of each circuit remains unchanged, there is

$$f = \frac{\partial W_m}{\partial g} |_{i=\text{const}} \tag{6.17}$$

In this formula, f is the generalized force and g is the generalized displacement.

If the generalized displacement is regarded as the mechanical angle in which the rotor excitation winding preceding the stator winding in reference to the axis, the generalized force can be regarded as the electromagnetic torque, then

$$T_{em} = \frac{\partial W_m}{\partial \gamma}\bigg|_{i=\text{const}} = \frac{\partial(\frac{1}{2}I^T LI)}{\partial \gamma}\bigg|_{i=\text{const}} = \frac{1}{2}I^T \frac{\partial L}{\partial \gamma} I \tag{6.17}$$

If the role pair number the motor is p, then

$$\gamma = \frac{\theta}{p} \tag{6.18}$$

The electromagnetic torque of the motor is:

$$T_{em} = \frac{\partial}{\partial \gamma}\left(\frac{1}{2}L_a i_a^2 + \frac{1}{2}L_f i_f^2 - Mi_f i_a \cos\theta\right) = pMi_a i_f \sin\theta \tag{6.19}$$

The motion equation of the rotor's motion is:

$$J\frac{d\Omega}{dt} = -T_{em} - T_o \tag{6.20}$$

Among them, T_o is the resistance torque, J is the inertia of rotation, Ω is the rotor's mechanical angular velocity, $\Omega = \frac{d\gamma}{dt}$.

Integrating the above formulae, the mathematical model of the pulsed alternator is,

$$\begin{cases} u_a = -L_a\frac{di_a}{dt} + \frac{d}{dt}(Mi_f \cos\theta) - i_a R_a \\ u_f = L_f\frac{di_f}{dt} - \frac{d}{dt}(Mi_a \cos\theta) + i_f R_f \\ J\frac{d\Omega}{dt} = pMi_a i_f \sin\theta - T_o \\ \theta = p\int_0^t \Omega dt + \theta_o \end{cases} \tag{6.21}$$

6.2.2.3 Self-Excitation Conditions

The self-excitation scheme of the air core pulsed alternator adopts the self-motivation method of positive feedback, and the principle of the excitation process is shown in Fig. 6.17.

The excitation capacitor discharges to the rotor excitation winding, injects the seed current i, and produces the excitation magnetic field. This provided, the armature voltage is induced at the end of the stator winding, and then the armature voltage is introduced into the excitation winding through the rectifier circuit, forming a circuit, producing a larger excitation current, therefore motivating a stronger magnetic field, and inducing a higher armature voltage. In this way a positive feedback is formed, allowing the excitation winding current to increase exponentially to the desired value.

The discharging pulsed amplitude and the number of the turns of armature windings of the motor are generally designed to be very small, so that the self-inductance value of the corresponding motor armature winding is also very small. On the other

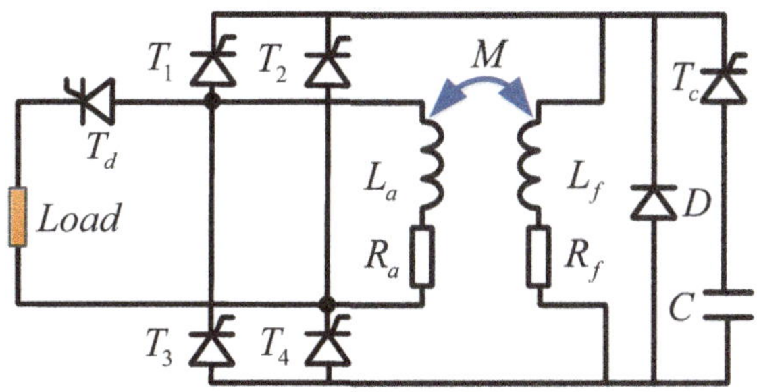

Fig. 6.17 Circuit topology of self-excitation

hand, in order to overcome the defects of the motor excitation difficulty caused by the low magnetic route conductivity of the air core pulsed alternator, the excitation winding of the motor is generally designed to be a winding with plenty of turns, so that the motor excitation winding's self-inductance value will be much larger than the armature winding. The value of their mutual inductance is somewhere in between.

The rectifier enters the motor excitation winding through the rectifier bridge to form an excitation current, so it can be considered that $|i_a| = |i_f| = |i_s|$. In the process of self-excitation, the self-inductance and resistance of the armature winding should be categorized to the rotor side of the motor and combine them with the inductance and resistance of rotor excitation winding, as shown in Fig. 6.18.

The voltage equation at the end of the armature is

$$\begin{aligned} u_a &= \frac{d}{dt}(Mi_s\cos\theta) = M\left[\cos\theta\frac{di_s}{dt} - i_s\sin\theta\frac{d\theta}{dt}\right] \\ &= M\left[\cos\theta\frac{di_s}{dt} - i_s\sin\theta(\omega + \frac{td\omega}{dt})\right] \end{aligned} \tag{6.22}$$

If ignoring the change in the rotation speed of the rotor, i.e.

$$\omega = \omega_{s_c}, \quad \frac{d\omega}{dt} = 0 \tag{6.23}$$

Therefore, the armature voltage can be simplified to

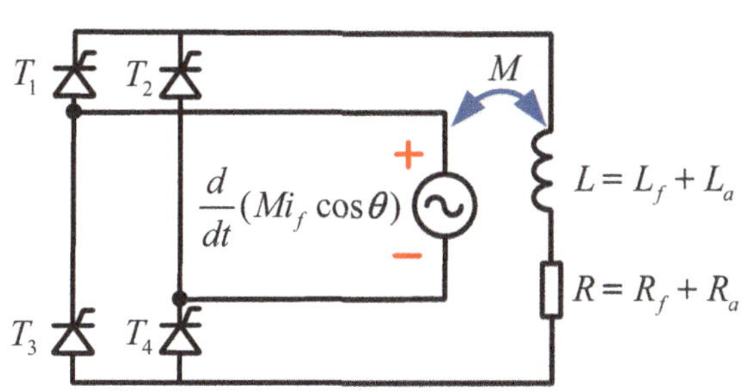

Fig. 6.18 The circuit diagram after calculation

$$u_a = M\left[\cos\theta\frac{di_s}{dt} - \omega_{s_c} i_s \sin\theta\right] = \sqrt{M^2\left(\frac{di_s}{dt}\right)^2 + (\omega_{s_c} M i_s)^2} \sin(\theta + \varphi_2) \tag{6.24}$$

$$U_a = \frac{2}{\pi}\sqrt{M^2\left(\frac{di_s}{dt}\right)^2 + (\omega M i_s)^2} \tag{6.25}$$

During the self-excitation process, the armature voltage is equal to the voltage of the excitation winding,

$$\frac{2}{\pi}\sqrt{M^2\left(\frac{di_s}{dt}\right)^2 + (\omega M i_s)^2} = L\frac{di_s}{dt} + Ri_s \tag{6.26}$$

$$\left(\frac{4}{\pi^2}M^2 - L^2\right)\left(\frac{di_s/dt}{i_s}\right)^2 - 2LR\frac{di_s/dt}{i_s} + \frac{4}{\pi^2}\omega^2 M^2 - R^2 = 0 \tag{6.27}$$

$$\frac{di_s/dt}{i_s} = \frac{-LR \pm \sqrt{R^2L^2 + (L^2 - \frac{4}{\pi^2}M^2)(\frac{4}{\pi^2}\omega^2 M^2 - R^2)}}{L^2 - \frac{4}{\pi^2}M^2} \tag{6.28}$$

If $\varepsilon = \frac{-LR \pm \sqrt{R^2L^2 + (L^2 - \frac{4}{\pi^2}M^2)(\frac{4}{\pi^2}\omega^2 M^2 - R^2)}}{L^2 - \frac{4}{\pi^2}M^2}$, the excitation current can be written as

$$i_s = I_o e^{\varepsilon t} \tag{6.29}$$

For ε, if the numerator and denominator are divided by L,

$$\varepsilon = \frac{-R \pm \sqrt{R^2 + \left(1 - \frac{4M^2}{\pi^2 L^2}\right)\left(\frac{4}{\pi^2}\omega^2 M^2 - R^2\right)}}{\left(1 - \frac{4M^2}{\pi^2 L^2}\right)L} \tag{6.30}$$

Because of $M^2 << L^2$, $\frac{k^2 M^2}{L^2} \approx 0$, ε can be simplified to

$$\varepsilon = \frac{-R \pm \frac{2}{\pi}\omega M}{L} \tag{6.31}$$

Therefore, the excitation current can be written as

$$i_s = I_o e^{(-R \pm \frac{2}{\pi}\omega M)t/L} \tag{6.32}$$

Excitation currents grow exponentially,

$$-R + \frac{2}{\pi}\omega M > 0 \tag{6.33}$$

$$\omega M > \frac{\pi R}{2} \tag{6.34}$$

From the above formula it can be seen that the higher the motor speed is, and the larger the maximum coupling mutual inductance coefficient between the stator and rotor is, the smaller the winding resistance will be, the easier the exciting starting will be, the faster the excitation current will increase, and the shorter the excitation time will last for. Moreover, because of the realization of high-power devices in the system, for the pulsed thyristor assembly in serial and paralleled forms, only a high positive voltage can ensure its reliability of recanalizing. If the rotation speed of the motor is too low, the induction voltage at the armature end is too small, there may occur a failure of the start exciting because the rectifier switch cannot be guided. In other words, the rising motor speed at the starting point can increase the output voltage at the end of the armature, thus ensuring the reliable recanalizing of the switching device in the rectifier circuit, and helping the electrical circuit to form a positive feedback.

Figure 6.19 shows the excitation current waveform during the motor's self-excitation process. The excitation current is $I_{excitation}$, and the seed current is I_o.

$$t_{excitation} = \frac{\ln(I_{excitation}/I_o)}{\varepsilon} \tag{6.35}$$

Ignoring the effects of the excitation winding resistance, assuming that the electrical energy stored in the excitation capacitor is fully converted to the rotor's inductor, there is

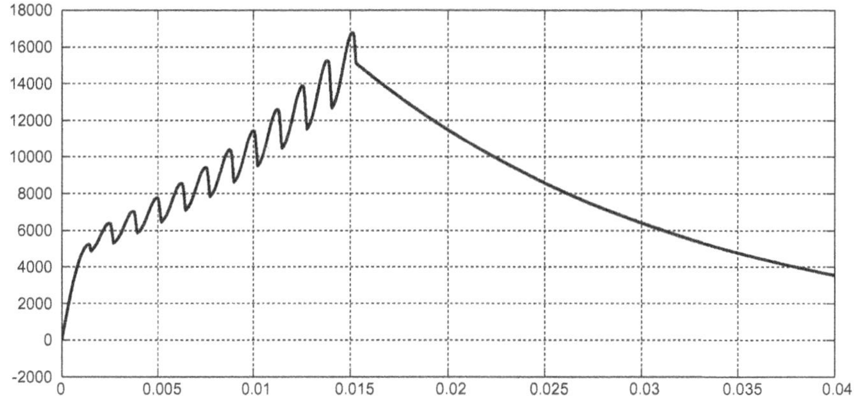

Fig. 6.19 Excitation current

$$\frac{1}{2}L_f I_o^2 = \frac{1}{2}CU_o^2 = E_o \tag{6.36}$$

The seed current provided by the reinforcement capacitor is

$$I_o = \sqrt{\frac{2E_o}{L_f}} \tag{6.37}$$

In summary, changing the motor design parameters to increase ε, and increasing the seed current I_o, that is, increasing the initial energy storage of the excitation capacitor, can both shorten the time required for the excitation process.

6.3 Discharging Process

6.3.1 High-Voltage Capacitors Provide Excitation Magnetic Field Energy

6.3.1.1 Single Pulse Excitation

When the load requires multiple pulses, the main switch for discharging can be triggered for multiple times, and the four discharging waveforms are as shown in Fig. 6.20. As shown in the Fig. 6.21 is the change of excitation winding current during discharge, and Fig. 6.22 is the change of rotation speed during discharge. As can be seen from the figure, the excitation scheme provides multiple continuous current pulses, and the amplitude of the pulsed current is reduced one by one. The speed of

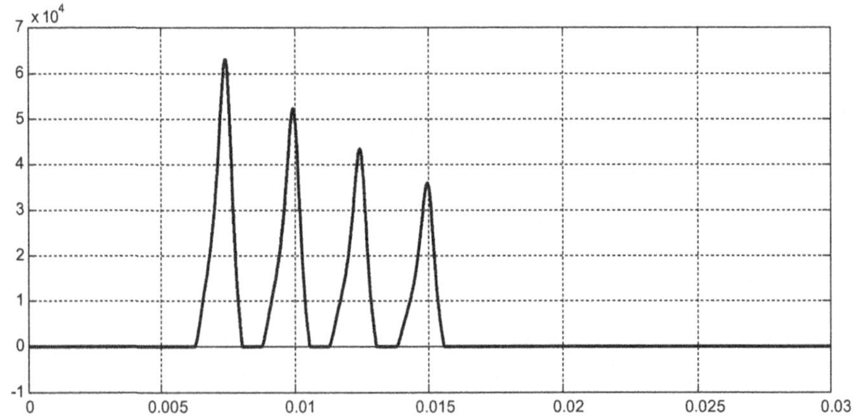

Fig. 6.20 Discharging waveform

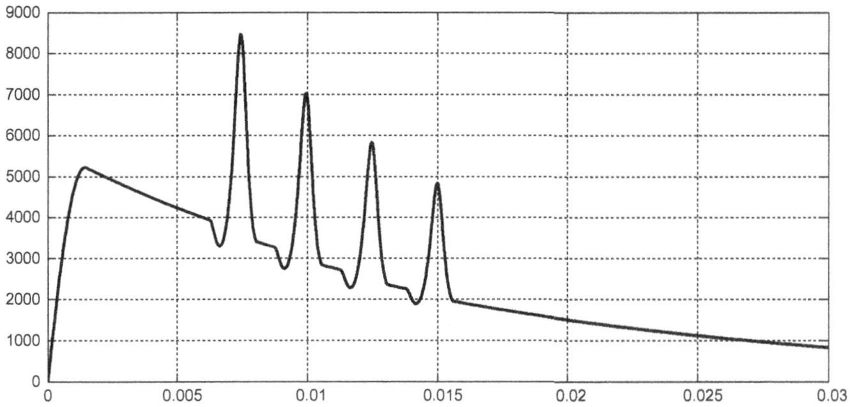

Fig. 6.21 Excitation winding waveform

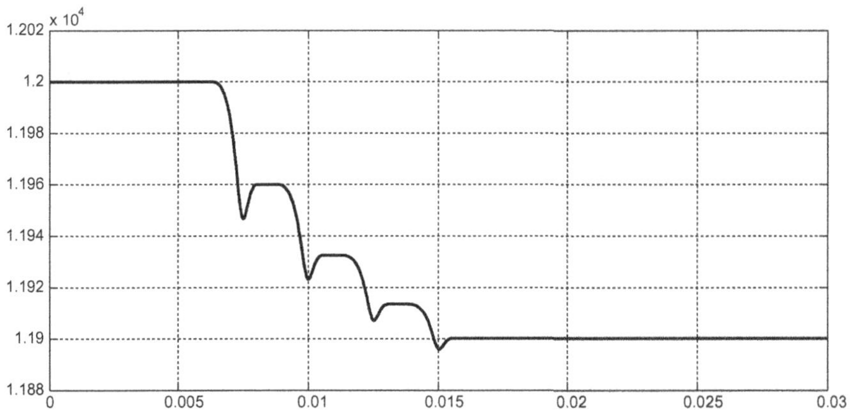

Fig. 6.22 Changes in speed during discharging process

the generator drops sharply through each pulse, indicating that the mechanical energy of the motor rotor is gradually converted into electrical energy. The magnitude of the pulsed current decreases one by one. The first reason is that the excitation current that the capacitor can provide is decreasing, and the second reason is that the generator's machinery is constantly converted into electrical energy, meanwhile the speed of the generator is decreasing.

6.3.1.2 Multi-Pulse Excitation

The motor rotating speed decreases step by step with each discharging process, and the amount of reduction increases each time. The excitation current is oriented after

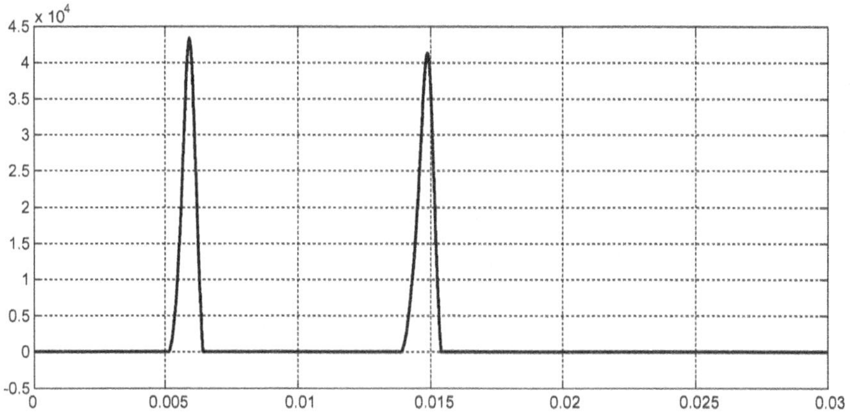

Fig. 6.23 Discharging waveform

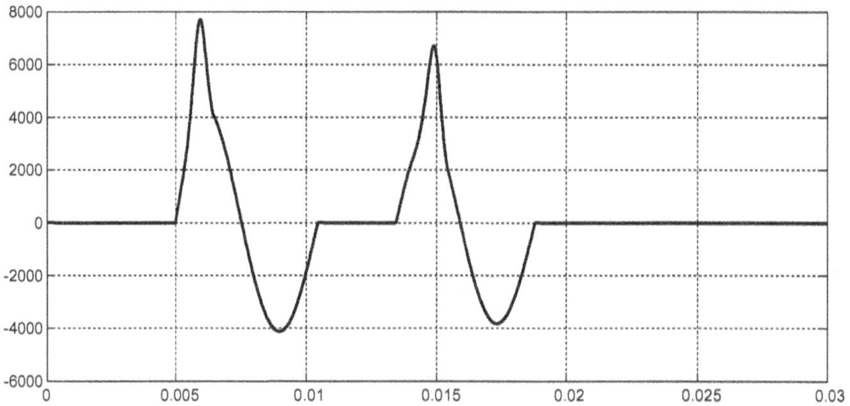

Fig. 6.24 Excitation winding waveform

each discharge, and there is a spike in each pulsed waveform of the excitation current, which is caused by the armature reaction. The armature current consists of a string of pulses whose magnitudes are basically equal to each other and the waveforms are basically consistent (Figs. 6.23, 6.24 and 6.25).

6.3.2 *Self-Excitation*

Since the start of the excitation process, the excitation current, excitation voltage and armature voltage of the amplitude continue to increase, while the rotating speed of the motor decreases at a certain rate, and the mechanical energy are converted into

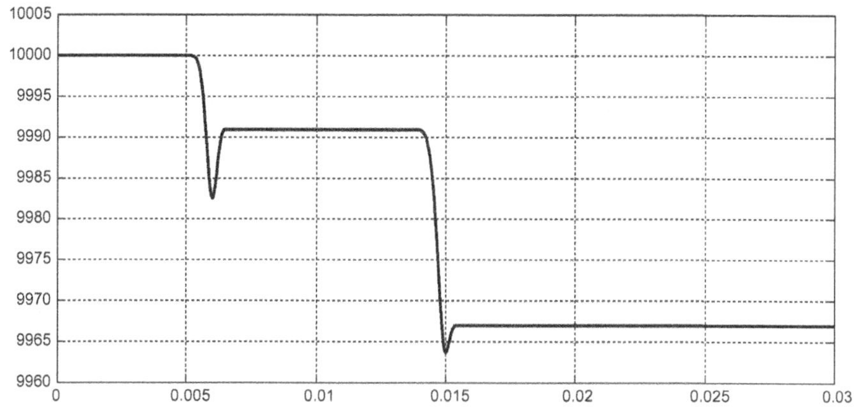

Fig. 6.25 Motor rotating speed curve

electromagnetic energy in the two sets of winding. When reaching the highest point, the armature circuit begins to discharge, the speed drops steeply, and there appears a high-value pulse in the armature waveform. After comparison, it can be found that the excitation current and armature current of the self-excitation scheme are higher than separate excitation (Figs. 6.26, 6.27 and 6.28).

However, the self-excitation scheme also has certain disadvantages. The establishment of its excitation requires a long time, its large-value current imposes a great impact on the service life of motor winding, brush, slip ring. In addition, self-excitation positive feedback requires strict monitoring in order to prevent the voltage and current to be an unneglectable damage to the motor and set fairly high bars for the controlling circuit and protection circuit.

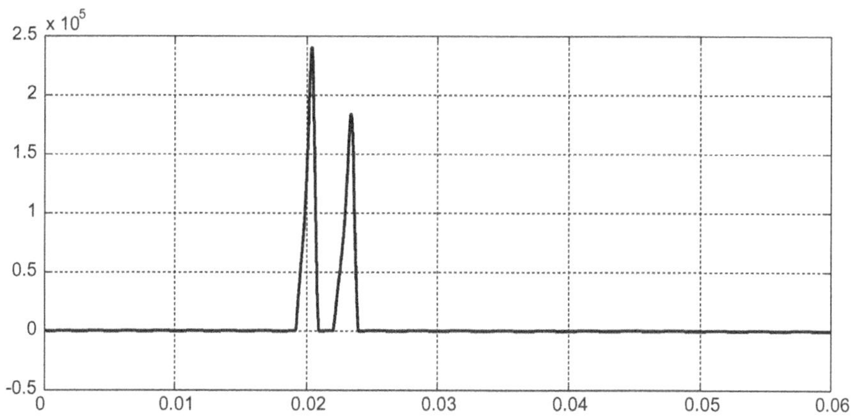

Fig. 6.26 Discharging waveform

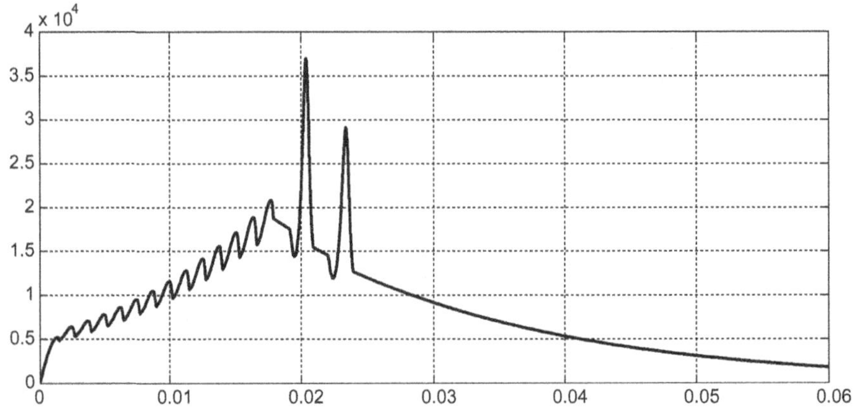

Fig. 6.27 Excitation current waveform

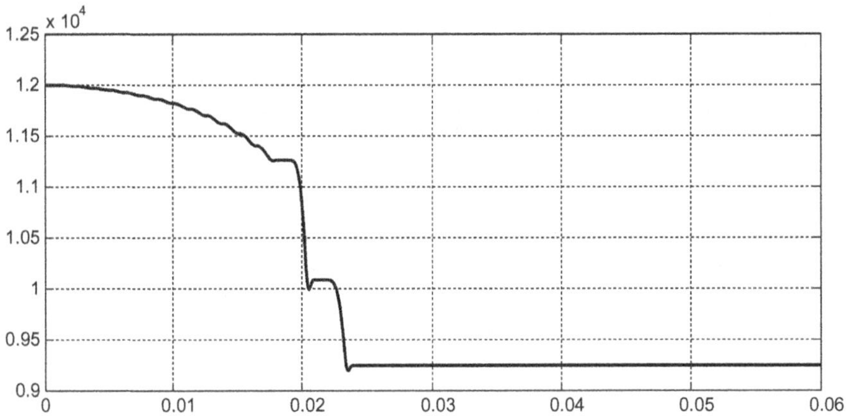

Fig. 6.28 Rotating speed curve

6.4 Energy Recovery

After the discharging of the pulsed alternator, there is still a large excitation magnetic field energy in the excitation winding. If the energy can be directly converted into heat energy and be discharged, it will not only reduce the overall efficiency of the system, but also make the heating problem more serious. During the after flow process after discharging, if the current in the excitation winding can flow through the armature winding, and by controlling the firing angle of the rectifier bridge, the system can be operated in an inverter mode, the electromagnetic energy in the excitation winding can be converted into the kinetic energy of the rotor, thereby improving the overall efficiency of the system.

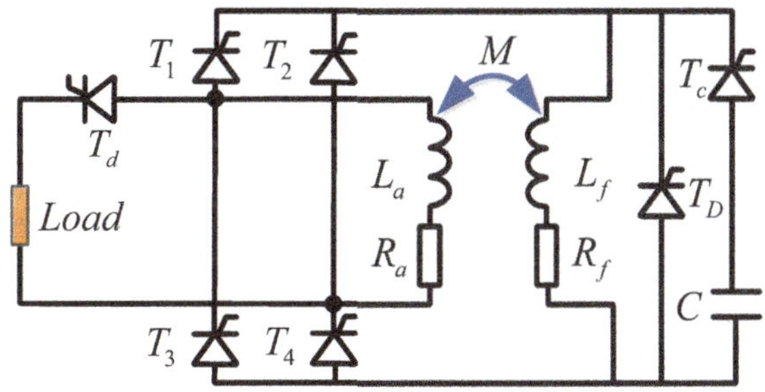

Fig. 6.29 Circuit schematic with energy recovery

In the previous self-excitation circuit topology, because of the use of continuous flow diode or diode bridge arm, the excitation winding cannot withstand the opposite voltage with the excitation current, and it's not possible for the electromagnetic energy be turned into negative power of the rotor kinetic energy. Hence it's hard to achieve the recovery of excitation current. Figure 6.29 shows a circuit schematic that enables excitation current energy recovery, in which the continuous flow thyristor T_D is in a state of shutdown during the process of self-excitation. When the excitation current reaches the rated value, it's required to close the self-excitation bridge, open T_D, in this way entering the phase of after flow, and at the same time, it's necessary to detect the position of the motor. When the discharging requirements are met, it's required to trigger the discharging switch of the main circuit. When the discharging is over, the self-excitation rectifier bridge is triggered again, hereby stopping applying the trigger signal to T_D, and T_D will be subject to reverse voltage and naturally shut down. When the current flowing through T_D is nil, the trigger angle of the excitation rectifier bridge is

$$\frac{\pi}{2} < \alpha < \alpha\mathrm{max}$$
$$\alpha\mathrm{max} = \pi - \beta_{\mathrm{min}}. \tag{6.38}$$

During this process, the system works in an active inverter state. When $\alpha_{>}\alpha_{\max}$, the inversion failed, and the electromagnetic energy in the excitation winding could not be recovered in the form of rotor kinetic energy. $\beta_{\min}$ is the minimum inverter angle. It is determined by the shutdown time, overlap angle and safety allowance angle of the pulsed thyristor assembly (Figs. 6.30, 6.31, 6.32 and 6.33).

The figure above shows the current and speed waveforms corresponding to the different inverter trigger angles, and the data that intercepts its key points is listed in Table 6.1.

As the chart above shows, as the inverse trigger angle gets closer to the excitation winding, the current decreases more and more rapidly during the energy recovery phase, and the speed rebounds more and more. When the inverse trigger angle exceeds $\alpha_{\max}$, the inversion fails, the current in the excitation winding cannot be recovered, and the rotation speed does not climb back.

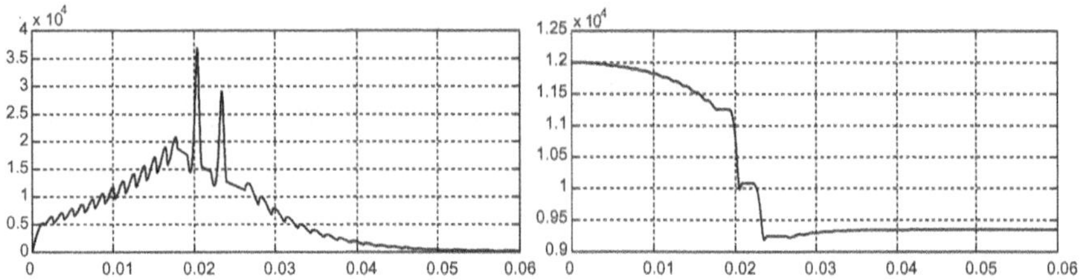

Fig. 6.30 Inverted trigger angle (130°)

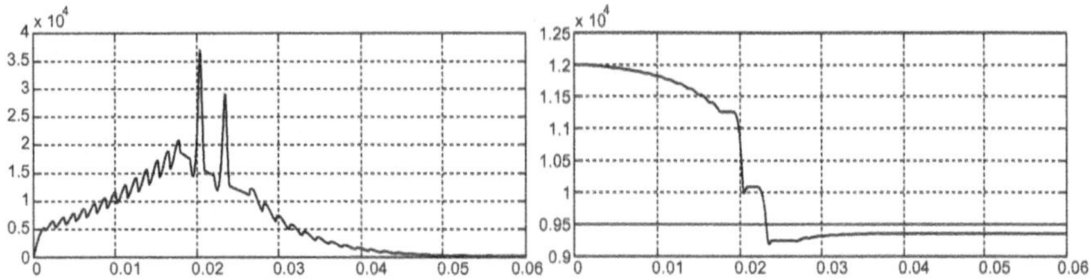

Fig. 6.31 Inverted trigger angle (140°)

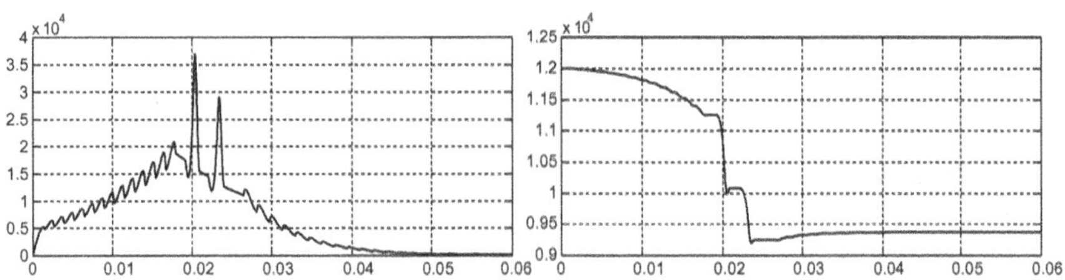

Fig. 6.32 Inverted trigger angle (150°)

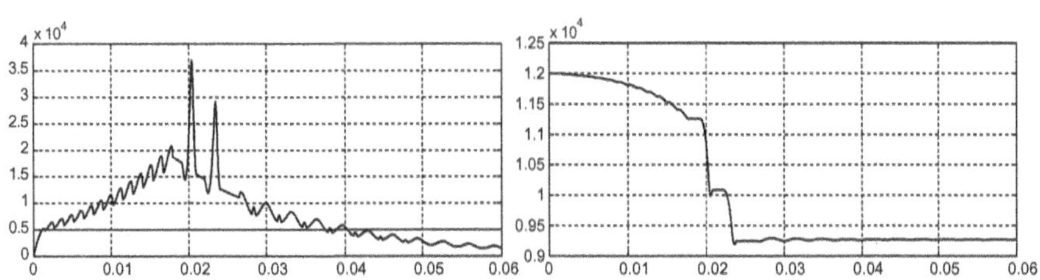

Fig. 6.33 Inverted trigger angle (151°)

6.5 Safety of the Pulsed Alternator

Figure 6.34 shows a diagram of the motor driving the self-excitation pulsed alternator's measurement and control system, which works as follows:

Table 6.1 Current and speed corresponding to different trigger angles

Excitation current (A)	0.03 s	0.04 s	0.05 s	0.06 s
130°	7872	1885	412	196
140°	7453	1586	306	143
150°	7174	1400	244	111
151°	10,060	5550	2900	1490
Without recovery	9180	5353	3118	1813
Rotation speed (rpm)	0.03 s	0.04 s	0.05 s	0.06 s
130°	9302	9348	9350	9350
140°	9315	9358	9360	9360
150°	9324	9365	9366	9367
151°	9252	9265	9269	9270
Without recovery	9247	9247	9247	9247

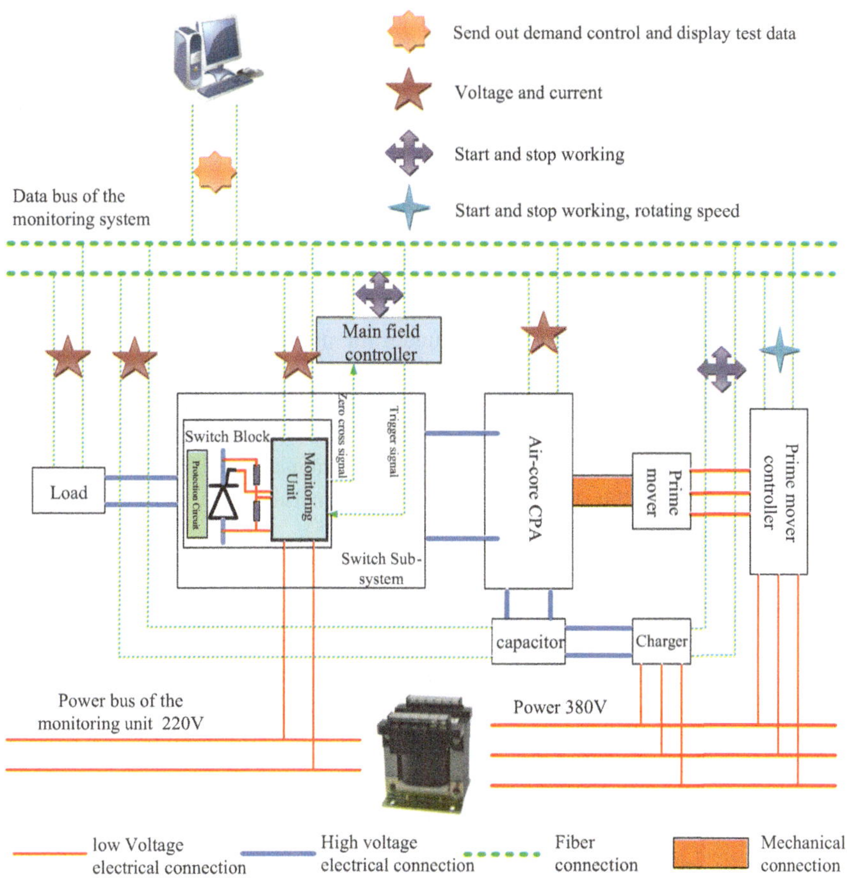

Fig. 6.34 A schematic of the inertial energy storage pulsed power control system

(1) The host computer issues a start-up instruction, each subsystem enters standby, all data transmitted by the sensor is returned to the host via the data bus.
(2) The capacitor charger charges the driving capacitor and stops when it reaches the rated value.
(3) The motivation controller drives the prime mover to the rated speed, and when the rotor reaches the rated speed and confirms that the other subsystems are working properly, the next process is handled by the field master controller.
(4) The field master first triggers the power switch of the initial excitation module, and the excitation capacitor discharges to the excitation winding on the rotor by a brush. The seed current will cause the armature winding to produce a smaller voltage. Since the armature windings are connected by excitation windings via a rectifier bridge, voltage and current increase exponentially, and the rotor's mechanical energy serves as the primary energy of this process.
(5) The voltage of the excitation current and the armature winding reaches the rated value and the opening enters the discharging process.
(6) After a predetermined period of self-excitation, the excitation current reaches the rated value, the field master triggers the discharging rectifier bridge in the predetermined discharge angle, and the load begins to discharge.
(7) The load is discharged by the current through the armature flow. When the scheduled discharging time is reached, the trigger ingress to the discharging bridge is stopped and the discharging is completed.
(8) Finally, the prime mover is controlled by the prime mover controller, so that the rotor is braked.
(9) Each sensor stops sending data to the host computer, reconstructs the waveform, and analyzes the data.

The above is the monitoring process under the normal operation of the inertial energy storage pulsed power supply. In addition, the failures probably also require to be considered. The design of the instantaneous power of the air core pulsed alternator can reach the 100 MW level. The amplitude of the pulsed current emitted by the motor is in the 100 kA level. The speed reaches tens of thousands of revolutions. These indicators are beyond the operating range of the general motor, so the system has high requirements for reliability and safety (Fig. 6.35).

(1) In the excitation phase, when the excitation rectifier bridge fails and cannot be closed, the current of the excitation winding will continue to increase. It needs to be determined through the current sensor of the excitation circuit. When the excitation winding current reaches a certain value, it's determined to be the excitation rectifier bridge failure, therefore the opening switch of the excitation winding is triggered, the circuit of the excitation winding current is cut off. At the same time the short circuit switch of the flow discharging circuit is triggered, and the current in the excitation winding is released.
(2) In the process of self-excitation, if the current sensor of the discharge circuit collects a current greater than a certain value, it indicates that the discharge switch is faulty, and the rectifier bridge needs to be shut down. The open switch of the discharging circuit is hereby triggered, and the discharging circuit is cut

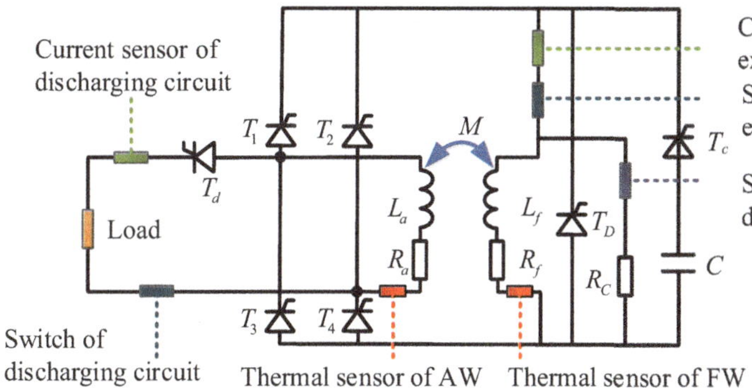

Fig. 6.35 System diagram

off in this way. The short circuit switch of the recovery discharging circuit is triggered, and the current in the excitation winding is released.

(3) Throughout the whole operating process, real-time monitoring of the armature winding and excitation winding temperature is carried out. When the rated value is exceeded, it is determined to be a fault, the whole system will hereby stop working.

Chapter 7
Electromagnetic Weapon Load of Pulsed Power Supply

Electromagnetic weapon is a new super-high-speed transmitter that uses electric energy to provide thrust for projectiles, including electromagnetic gun and electrothermal gun. The electromagnetic gun is a device that uses electromagnetic force to push projectiles, and the electrothermal gun is a device that uses electric energy to heat the working fluid and push the projectile. Compared with traditional guns, it boasts high projectile speed, simple control and energy request, and easy concealment, so it has a wide range of applications in military.

7.1 Railguns

Railgun is the highest-standard, fastest-growing, most invested and most promising advanced technology weapon in the electromagnetic weapons project. It is expected to be officially equipped with ships in the next few years.

7.1.1 Fundamental Principles

Railgun is a very simple electromagnetic transmitters consisting of a paralleled metal guide rail, an armature, projectiles, and a pulsed power supply, as shown in Fig. 7.1. The armature is located between two rails and is composed of conductive material. The two metal guide rails are also good conductors. The materials should have better mechanical strength in order to withstand ablation and friction. The orbits are often embedded in high-strength composite insulators for the formation of gun tubes. In addition to conducting high currents, the guide rails are also used to guide the armature to make the projectile move.

S. Wu and S. Cui, *Pulsed Alternators Technologies and Application*,
https://doi.org/10.1007/978-981-33-4224-8_7

Fig. 7.1 Schema of railguns

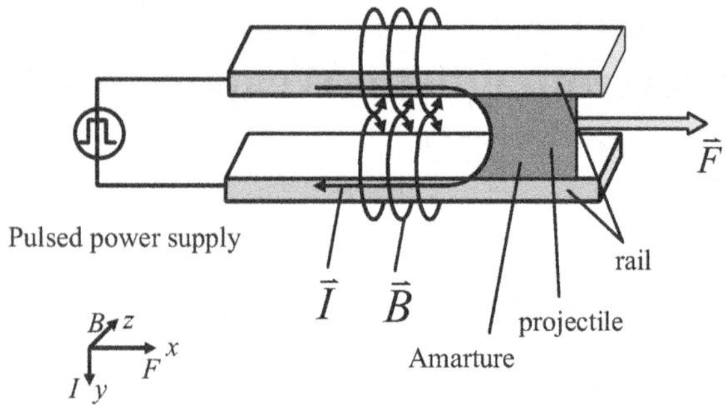

When the pulsed power supply is triggered by the control circuit to provide a strong current pulse for the railgun, the current passes through the upper, armature, and lower rails and therefore a circuit is formed. As is known from the Faraday's law of electromagnetic induction, when the current flows through the guide rail, a magnetic field is generated in the area between the two rails and the armature, which is placed in the magnetic field and influenced by the electromagnetic force that propels the armature forward to accelerate the motion.

In order to understand the nature of the railgun better, it can be regarded as a port electromechanical energy transformation device. Therefore, by the principle of electromechanical energy transformation, it can be seen that the energy of the railgun is:

$$W = W_m + W_k = \frac{1}{2}L_r I^2 + \frac{1}{2}mv^2 \tag{7.1}$$

W_m the magnetic energy of the railgun system;
W_k, The kinetic energy of the armature and projectiles.

The energy variation rate of the railgun system is:

$$\frac{dW}{dt} = L_r \frac{dI}{dt} + \frac{1}{2}\frac{dL_r}{dt}I^2 + mv\frac{dv}{dt} \tag{7.2}$$

The electrical power of the input railgun system is:

$$P_e = VI = I\left(L\frac{dI}{dt} + I\frac{dL}{dt}\right) = I\left(L\frac{dI}{dt} + IL'v\right) \tag{7.3}$$

where

L' inductive gradient of the orbit.
$\frac{dW}{dt} = P$, the electromagnetic force acting on the armature is

$$F = ma = \frac{1}{2}L'I^2 \tag{7.4}$$

Through this equation, the mechanical quantity and electrical quantity can be linked to investigate the load characteristics of the railgun.

7.1.2 Load Characteristics

Railguns are composed of orbits and armatures, in which the orbit is composed of conductive metals, with a certain degree of impedance. When the projectiles continue to move forward, the length of the rail in the launching circuit continues to grow, the impedance of the rail also increases, so the railgun can be equivalent to a time-varying resistance and a time-varying inductive circuit. The law of change is related to the position of the projectile in the railgun and is bound by the circuit and the motion equation.

The rail's equivalent resistance R_x can be expressed as:

$$R_x = R_{x0} + R'x \tag{7.5}$$

R_{x0} the initial resistance of the rail in Ω;
R' the resistance gradient of the rail in Ω/m;
x the displacement of projectiles in the railgun in m.

The resistance gradient of the rail, R', is a complex function controlled by multiple variables, under the skin effects of the velocity to the cross-sectional area of the carrier and the effect of the high temperature phase change caused by the friction between armature and orbit. It can be considered to be constant when calculated by simplified model,

The rail's equivalent inductor L_x can be expressed as:

$$L_x = L_{x0} + L'x \tag{7.6}$$

L_{x0} the initial inductor of the rail in H;
L' Inductive gradient of the rail in H/m.

The inductive gradient L' of the rail is related to the structure and size of the rail, and the inductive gradient can still be considered to be a constant value when a simplified model is used.

The displacement x of the projectile is related to the acceleration process. The projectile is exposed to the joint effect of four forces in the railgun's acceleration process, they are: electromagnetic propulsion F_L, air resistance F_p, friction resistance F_f, and ablation resistance F_{ab}. According to Newton's second law, its motion equation is:

$$ma = F_L - F_p - F_f - F_{ab} \tag{7.7}$$

m the quality of projectiles in kg;
a acceleration of projectiles in m/s^2.

In the formula (7.7), air resistance F_p, friction resistance F_f, ablation resistance F_{ab} can all be expressed as a function controlled by multiple variables such as railgun structure, current, velocity, etc. Electromagnetic propulsion F_L is only to be considered in the ideal railgun model. In this way the projectile acceleration can be expressed as:

$$a = \frac{F_L}{m} = \frac{L' i_L^2}{2m} \tag{7.8}$$

i_L railgun input current in A;
a acceleration of projectiles in m/s^2.

The speed v and displacement x of the projectile can be expressed as:

$$v = \int_0^t a dt \tag{7.9}$$

$$x = \int_0^t v dt \tag{7.10}$$

The acceleration process of visible projectiles is related to the input current of the railgun, so the discharging circuit current needs to be analyzed.

The circuit simplification model of the pulsed driven railgun system is shown as Fig. 7.2, which combines the inductor and resistance on the line with the initial inductor L_{x0} and the resistance R_{x0} in the railgun to be L_{line} and R_{line}. They are

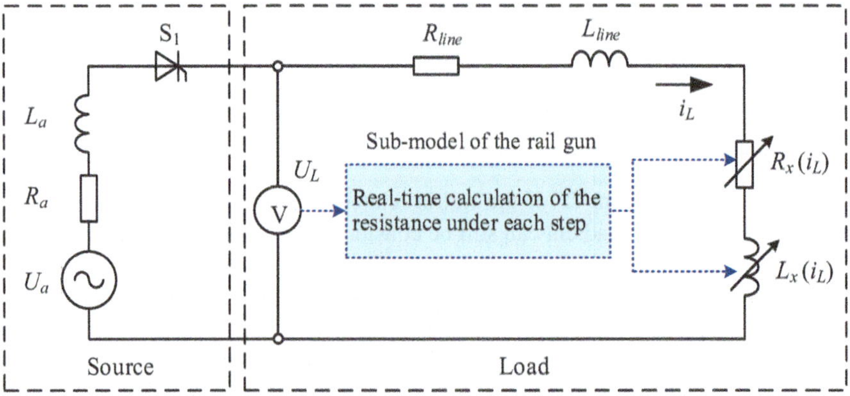

Fig. 7.2 A simplified circuit model of a pulsed alternator-driven railgun

considered to be parts of the load model on the side of the railgun. The mathematical model and finite element model established earlier can therefore be simulated.

According to the formula (7.4) to (7.10), the railgun's equivalent resistance R_x and inductance L_x can be expressed as function of the load current i_L, and the voltage equation of the return circuit equation can be written and the connection between the load side and the power side can be established.

According to Kirchhoff's law, voltage equation on the load side is

$$U_L = (R_{line} + R_x)i_L + L_{line}\frac{di_L}{dt} + \left(L_x\frac{di_L}{dt} + i_L\frac{dL_x}{dt}\right) \tag{7.11}$$

Therefore, according to the above equation, a simplified model of the railgun can be built in the simulation software, and the required pulsed power supply can be easily estimated.

7.1.3 Key Technical Issues

From the research progress, the electromagnetic railgun is likely to become the first electromagnetic weapon to realize weaponization because of its simple structure and rapid development, but there are still some key technical problems to be solved.

7.1.3.1 Anti-Ablation Technology

The representative key technology of electromagnetic railgun is the guide rail ablation resistance technology. The results show that the ablation on the surface of the orbital electrode through which the projectile passes is more serious. The shapes of the surface of the orbital electrode after ablation are mainly the arc marks, the craters and the droplet particles suspended on the upper orbit electrode. Ablation not only seriously affects the life of the injection duct, but also the presence of attached metal fragments before the projectile, which may cause the failure of the launch.

Ablation mainly arises from several reasons: in the initial acceleration stage, the current change is very large, the change of electromagnetic force produces a great oscillation, here comes the crater of the starting segment; because of the low armature resistance rate and the ohmic thermal dissipation, it's very easy for the loss of contact with the guide rail, which causes the ablation. However, this is not a problem of plasma, even high resistance rate, high voltage drops and the easiness for arc production make the emission efficiency very low. What is more serious is that the arc will produce serious ablation.

In order to reduce ablation on the orbital surface, a variety of studies have been carried out, such as multi-stage and segmented orbital structures, distribution energy storage to regulate current waveforms and composite technology. In addition, the United States is currently developing a solid armature for electromagnetic railguns, which is made of metal materials that are heavier than plasma armatures and are slower than they are. However, the material does not produce plasma violently, and it is more practical than a plasma armature when it comes to metal contact. It is assumed that the speed of about 3 km/s can be obtained by using solid armature, and its application prospect is still to be further explored in accordance with the actual conditions of use.

7.1.3.2 Continuous Firing Technology

So far, indoor tests have focused on the development of electromagnetic acceleration technology, so there is no need to develop continuous firing technology. The main problem of continuous firing technology is the thermal management of transmitters and power supplies. When the power supply is charged and discharged repeatedly at high speed, the internal heating of the capacitor may cause deterioration, whose performance and parameters must be determined by test.

In addition, in the continuous firing, the surface of the rail may be covered by molten armature material. Before the material cures the next shot, whether the molten metal plays a lubricant role, and whatever the impact on the second shot will be after curing depend on the future tests.

7.1.3.3 Emission Structure Technology

The shells fired by electromagnetic guns consist of armatures, butts and projectiles. Because the projectiles are very fast at first and the kinetic energy is very high, they are usually not equipped with explosives, but rely entirely on kinetic energy to destroy the target. At present, the focus of projectile research includes: the acceleration of projectiles on the rails, the compatibility of the projectile assembly with the rail transmitter, and the lighter, stronger projectile material, reducing the ablation of the inner wall of the transmitter.

The armature is one of the key components of the railgun. The armatures selected in current study include plasma armature and solid armature. The solid armature is the focus of current and future research, and also the main point of electromagnetic gun engineering application in the future. Early research by the University of Texas System of Advanced Technology focused on the role of geometry in the distribution of current density and temperature. They also designed the saddle armature based on the C-armature, thus reducing the uneven distribution of current density and thus controlling armature ablation. In addition, the guide-rail-type electromagnetic mortar's muzzle is equipped with a muzzle diverter, which can export the residual energy on the rail after projectile firing, enabling rapid silence and flame cancellation. The muzzle diverter mainly consists of an inductive muzzle diverter, which recovers the remaining energy through a cable connected to the muzzle to the tail of the gun, and a resistant muzzle diverter, which exports the remaining energy directly.

The most distinguished feature of an electromagnetic railgun is the ultra-fast speed of the projectiles fired. However, the atmospheric resistance to the projectile increases proportionally with the square of the velocity. Therefore, the projectile must be designed to be in an appropriate shape in order to take advantage of the ultra-high speed. Electromagnetic railguns that meet the requirements of actual equipment are not suitable for ultra-fast flight. Now, the initial speed of the projectile is aimed at 2500 m/s, however, the speed is already in the category of hydrodynamic hypersonic speed. The front end of the projectile will be heated violently when flying at that speed at atmospheric pressure. In long-distance flight, any small deformation will have a great impact on the flight orbit, also on the accuracy of the hit, so compared with the general artillery, the electromagnetic railgun requires precision of the mechanical components.

7.1.3.4 High Current Controlling Technology

The strong current emitted by the pulsed power supply cannot be used directly for electromagnetic emission, because the electromagnetic thrust generated by a single pulse dielectric is accompanied by strong oscillations and low energy conversion efficiency. It is found that the ideal external characteristic of the transmitting power supply current is the trapezoidal flattened pulse. The rapid rise in current allows the armature to gain a high level of electromagnetic thrust in a very short period of time. During acceleration, the flattened current provides a constant thrust. The small impact of the uniform acceleration to the rail reduces wear effectively. When the projectiles leave, the residual electrical energy may break through. Therefore, the orbital current should fall back rapidly after the launch.

At present, the common method is to discharge with multiple pulsed power asynchronously for a synthesis of approximate trapezoidal waves. In the case of continuous emission, the control and regulation of the repetitive charging, discharging and feed of the multiple pulsed power supply are the cores of power regulation.

Switch control of power supplies remains a major challenge for power supply technology due to high-power pulsed currents up to mega-amps. Such high currents and high voltages are well ahead the threshold for a single SCR, and only a SCR array can meet such electrical requirements. Additionally, the array is too large in volume and mass. Studies have shown that the use of SiC instead of Si in array switches, not only reduced volume and quality, but also significantly improved its performance. However, SiC is not commercially available. Therefore, it is particularly important to research and develop the large-area production.

7.2 Coilguns

Coilgun is an important part of electromagnetic gun. Compared with the railgun, it has not received much attention in a period of time, mainly because of its technical complexity. In recent years, due to the development of pulsed power technology, it is recognized that the inductive gradient of the railgun has nothing to do with the size of the gun, while inductive gradient has been greatly improved in the large-scale coilguns. In recent years, people are becoming more interested in coilguns again.

7.2.1 Basic Principles

There are more than a dozen types of coilguns, of which only the ones with synchronous induction and asynchronous induction are the two most commonly used categories.

7.2.1.1 Synchronous Induction Coilgun

The synchronous induction coilgun is a launch tube consisting of fixed driven coils. There is a projectile coil inside, which is fixed to the projectile, and the driven coil and the projectile coil are arranged in the same axis. The model is shown in Fig. 7.3. For the convenience of analysis, the driven coil and projectile coil are reduced to current rings C_1 and C_2. When a variable current i_d passes through the coil C_1, a changing magnetic field is generated around the coil. When the magnetic field changes on the cross-section of coil C_2, according to the law of electromagnetic induction, it can be seen that current i_p is flowing in C_2. Assuming the interaction between the

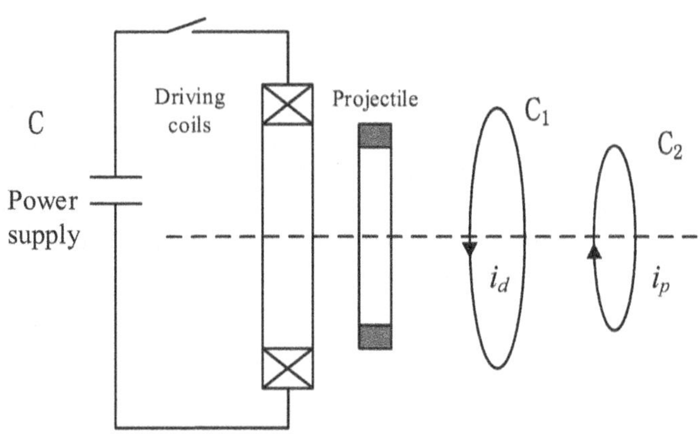

Fig. 7.3 The operation schematic of unipolar induction coilgun

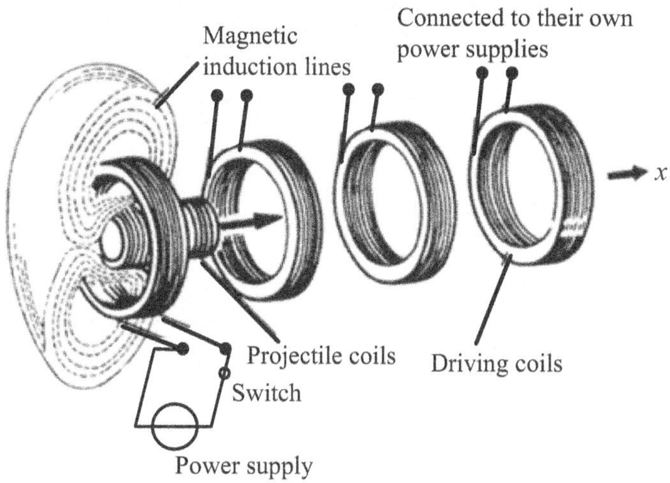

Fig. 7.4 The operation schematic of multipolar induction coilgun

projectile and the drive coil is M, it can be conducted that the driving force acting on the projectile is F_p, which propels the projectile forward, as shown in the formula (7.12).

$$F_p = i_d i_p \frac{dM}{dx} \tag{7.12}$$

Generally, to reduce the fluctuation of the force F_p and extend its accelerated travel, the above drive coils are made into a multi-stage structure, which is commonly referred to as multi-stage induction coilgun, whose working principle is as shown in Fig. 7.4. In multipolar induction coil, the projectiles are accelerated in turns, and finally reached ultra-high speed. The theoretical efficiency is expected to be 100%.

7.2.1.2 Asynchronous Induction Coilgun

The asynchronous induction coilgun consists of a driving coil and projectile, and its principle comes from the asynchronous induction motor. It is as shown in Fig. 7.5. The driving coil is connected into a continuous coil of a multiphase winding, stimulated by a multi-phased (mostly three-phase) power supply, which produces a linear wave magnetic field like the rotating magnetic field of the asynchronous motor. The wave velocity is faster than the projectile speed, causes relative motion with the help of its slip speed, induces the current (vortex) inside the projectiles, and "drives" the projectile forward by its magnetic field. As the speed required for pellet acceleration is increasing, the entire coil should be divided into a few segments. In order to obtain a line wave whose phase velocity increases from one phase to another, the frequency of the excitation power supply should be increased, or the pitch of the drive coil

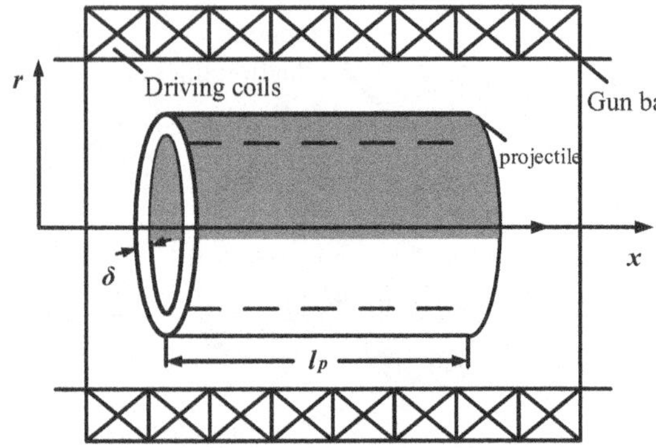

Fig. 7.5 The schematic of asynchronous induction coilguns

(half wavelength) should be increased. Because the projectile length is relatively short, excessive increase in the pole distance is not practical, it is more appropriate to increase the power supply frequency along the length of the gun barrel. It's possible to use one frequency per segment, only the piece-wisely increased frequency. Therefore, the excitation frequency of each segment of the asynchronous induction coilgun is different, so it can be used for different frequency generators or capacitor circuits with different resonant frequencies as the power supply for the asynchronous induction coilgun. The thrust F_p of the projectile is:

$$F_p = \frac{I_d^2 R_p X_m^2 \alpha}{v_w s\left[\left(R_p/s\right)^2 + X_m^2\right]} \tag{7.13}$$

I_d the pulse current of the driving coil;
α the ratio of the length of the projectile coil to the excitation part of the driving coil;
R_p projectile driving coil resistance;
X_m the electrical resistance of the excitation winding;

Slip $s = v_s/v_w = \left(v_w - v_p\right)/v_w = 1 - v_p/v_w$, among which v_w is the wave velocity of the wave magnetic field, the speed of the slip $v_s = v_w - v_p$;

v_p Projectile coil velocity.

7.2.2 Key Technical Issues

Considering the current situation and difficulties of asynchronous induction coil transmitter, its key technologies mainly include the following aspects: power supply technology, synchronous control technology, material technology and system technology.

(1) Power supply technology

Because asynchronous induction coil transmitters have high demands on the voltage and frequency of the supplies, and the energy is quite large, it is difficult for ordinary power supplies to meet the requirements. Typically, the power of the primary power supply is passed to the energy storage system initially, the energy is stored in the pulse-driven system, and then released in due course to meet the high-energy pulses we need. At present, the main power supply used by the engine model includes capacitor group, the flux compression generator of the inductive energy storage system, battery pack, pulsed magnetic fluid power generation device, unipolar pulsed alternator and compensating pulsed alternator. Each kind has its own characteristics and practical value. From the current research and test, the focus of the research is: capacitor banks with high energy and high energy storage density, unipolar generators, and compensating pulsed alternators. These kinds of power supply are developing fairly rapidly, and their application is becoming more mature. The challenge remaining is to reduce their sizes.

(2) Controlling problems of driving coils

Because the inductive emission requires a long-lasting magnetic travelling wave, the phase coil current should maintain a certain phase difference, and the discharge of the driving coil should be closely coordinated with the movement of the projectile, which puts more stringent requirements on the control of coil excitation.

(3) Material technology

Because the electromagnetic emission is accelerated under strong pulsed current, its working conditions are rather harsh, the requirements to the materials are relatively high. In particular, the temperature rise of the projectiles in asynchronous induction coil transmitter is pretty large, so is the acceleration forcer, therefore the bars for the projectile's high temperature resistance and mechanical strength are very high.

(4) Overall technology of the system

The research of electromagnetic emission is mainly about how to improve the speed of the emitter. Over the years, it has been carried out a lot of research on the relevant single technology and already made great progresses. When a single technology develops to a certain extent, the overall technology of the system becomes a very important key technology. From the aspects including overall situation of the system, the function of each component, the selected technical approach and the implementation plan, quantitative indicators and corresponding constraints for the research and development of each sub-system and components must be put forward, in order to obtain the optimization of the overall comprehensive performance of the system.

7.3 Electrothermal Chemical Guns

Electrothermal chemical gun is an electric emission technology. It is a typical combination of electro-physics and chemistry. Because the development of chemical gun by its own is limited by the inherent characteristics, its initial speed is no longer possible to improve. That means it can only be limited in a range of few hundred meters/second. Even for the ones with high pressure the initial speeds are no more than 1600 m/s. The initial speed of artillery is mainly limited by the structural strength and the large molecular weight of the gas used in the propellant. In order to improve the initial speed, from World War II, people began to use electropulsing heating to produce high-temperature high-pressure plasma, in an attempt to accelerate the projectile to a high speed by electricity. It is expected that the electrothermal chemical gun can accelerate the projectile to 2.5–3.0 km/s. So far, the prototype of electrothermal chemical gun has been successfully developed, even qualified for armies.

7.3.1 How Electrothermal Chemical Guns Work

The principle of electrothermal chemical gun is to use high-capacity, high-power pulsed power supply to implement high voltage, high current discharge to the plasma generator, produce high-temperature high-speed plasma ignition chemical refrigerants, whose combustion generates a large number of high-temperature high-pressure gas, which expands to promote the advance of the projectile. As shown in Fig. 7.6 is a diagram of the electrothermal chemical gun emission system, mainly composed of power supply system, intermediate energy storage part, pulsed forming network, switch, plasma generator, anther chamber, projectile and gun barrel.

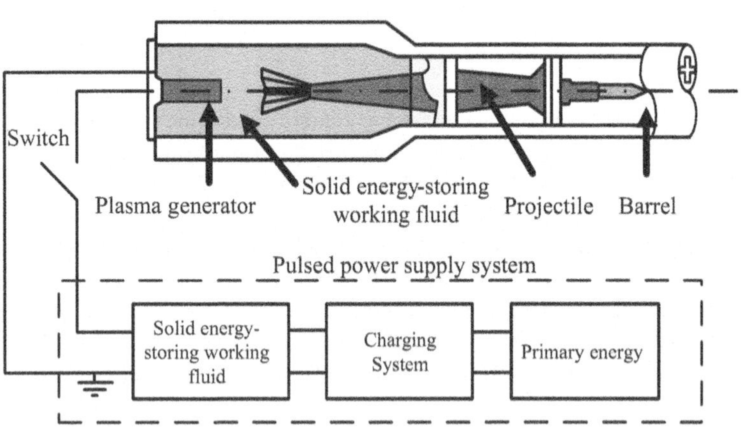

Fig. 7.6 Diagram of electrothermal chemical emission system

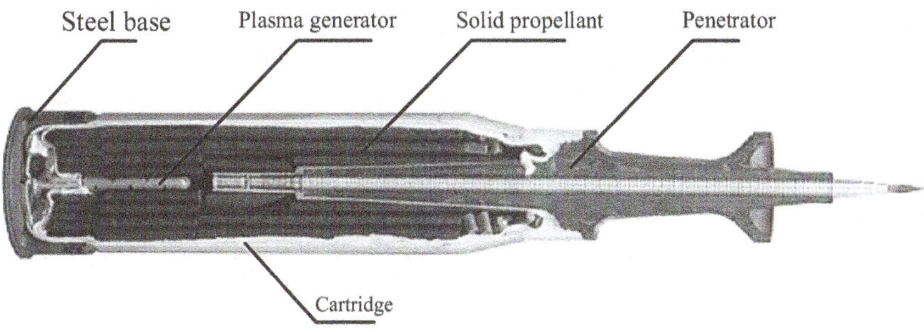

Fig. 7.7 Diagram of the structure of electrothermal chemical solid propellant gun

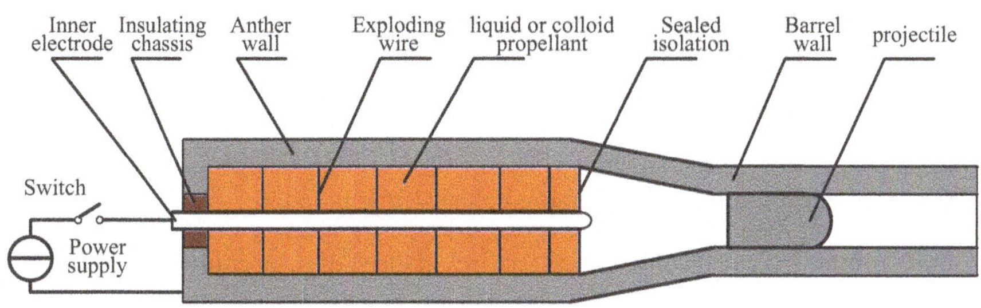

Fig. 7.8 Diagram of the structure of electrothermal chemical liquid-launching artillery installation

According to the form of the substances that contains energy after projectile, electrothermal chemical gun is generally divided into electrothermal chemical solid propellant gun (ETCSPG) and electrothermal chemical liquid propellant gun (ETCLPG). Figures 7.7 and 7.8 illustrate the diagrams of the structures of electrothermal chemical solid propellant gun and electro-thermochemical liquid propellant gun respectively.

7.3.2 Load Characteristics of Electrothermal Chemical Gun

In electrothermal chemical emission system, plasma generator is the main load form of pulsed power system discharge, which affects the output law of pulse-forming network power, and directly determines the energy utilization efficiency of power supply system.

Since the 1980s, extensive research has been carried out on the basis of the theory of the working characteristics of plasma generators at home and abroad, and a great deal of simulation research has been implemented on the discharge process from the plasma micro-model. These models explain the corresponding experimental phenomena very well, but the theoretical parameters of the model are very complex and require a lot of other parameters, therefore it's difficult to understand and analyze.

In addition, the acquisition and application of these parameters are also not easy to achieve. Taking the discharging circuits with centralized parameters into consideration, the capillary discharging plasma is treated as a time varying resistor R_h, whose resistance value is of complex nonlinear characteristics, which can be divided into the stage of fissure burst and the stage of capillary plasma arc. The resistance value of the previous stage is determined by the resistivity in the process of phase transition of the wire. The resistance of the latter stage is determined by the input current, capillary structure, plasma refrigerant, temperature and pressure, etc., which can be calculated by capillary arc resistance model, reducing the temperature, pressure and other factors to the coefficient of resistance:

$$R_h = 0.22lk^{\frac{8}{11}}a^{-\frac{13}{11}}i^{-\frac{6}{11}} \tag{7.14}$$

l capillary length, in cm;
k proportional constant, approximate to a constant for capillary tubes of a certain material;
a capillary section radius, in cm;
i the input current of the electrothermal chemical gun, in kA

By combining the resistance characteristics in the stages, an equivalent resistance of typical electrothermal gun can be obtained, as shown in Fig. 7.9. In this way, it can be told that in the arcing stage of the discharging process, the resistance rose rapidly for a short time, and reached hundreds of mmho immediately, and then entered the plasma discharging stage, the resistance value quickly dropped to only a few tens of mmho. During the maintenance of arc, the range of the change of the resistance is not large. When the voltage is not strong enough to maintain the arc, the current decays, and the resistance rises, until the arc extinguishes.

According to the above analysis, when building a joint simulation model of the pulse dynamo-driven electrochemical gun system on the simulation platform, the electrothermal chemical gun can be equivalent to a variable resistance, the output

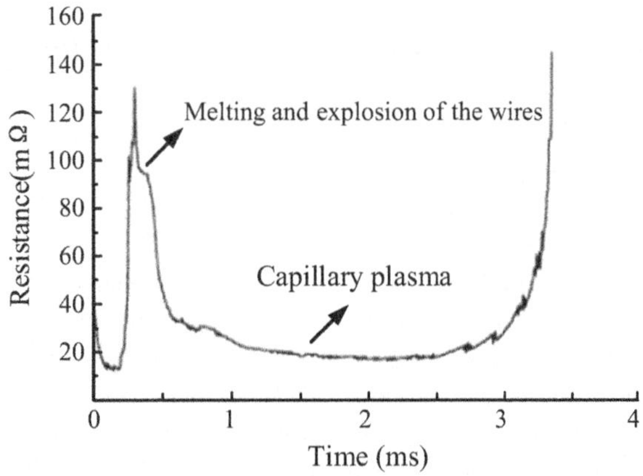

Fig. 7.9 Equivalent resistance to electrothermal chemical gun

current can be obtained with the initial value of the given resistance in simulation, the resistance value is calculated at the next moment as shown in formula (7.14), and then fed back to the circuit model to get a new output current. In this way, the simulation results of the entire discharge process are obtained through this cycle.

The time-varying resistance model of electrothermal chemical gun can also be used to check the tables. Based on the experimental data of the same type of electrothermal chemical gun, the equivalent resistance value of the plasma arc at a given time is brought directly into the system simulation model.

7.3.3 Prospects for Future Discovery and Key Technologies

Compared with railguns, electrothermal chemical guns require relatively less external energy. In contrast, electrothermal chemical gun firing technology is the most mature and possible in electromagnetic weapons, and it is expected that electrothermal chemical artillery will become the replacement of the current solid propellant guns.

In 2006, the U.S. Army's ETC firing experiment with light 120 mm XM-291 gun succeeded, using electrothermal chemical emission technology. The muzzle energy was very close to the 17 MJ, equivalent to the lower limit of the conventional one which is 140 mm. It is shown that ETC technology is feasible and can be integrated into existing gun systems.

The success of the XM-291 in the United States proves that it is feasible to continue the research of electrothermal chemical gun technology and integrate it into modern tanks. Although the test has been successful, there are still many problems to be solved, for example, in order to further improve the initial speed of the projectile, it is necessary to further explore the ignition of electrothermal chemical gun, so that it can be applied to liquid propellants. In addition, electrothermal chemical gun technology should be compatible with existing systems to reduce the rear-seat force passed to the vehicle when the gun is fired, which is essential for the integration of electrothermal chemical guns and chariots. In addition to the problems related to the assembly of the system remain unresolved, the main technical challenge of electrothermal chemical propulsion technology is the development of new high-energy, low-vulnerability propellant.

In the future, the development of electrothermal chemical gun emission technology will mainly focus on technology integration and engineering, and its technical difficulties mainly include the following aspects.

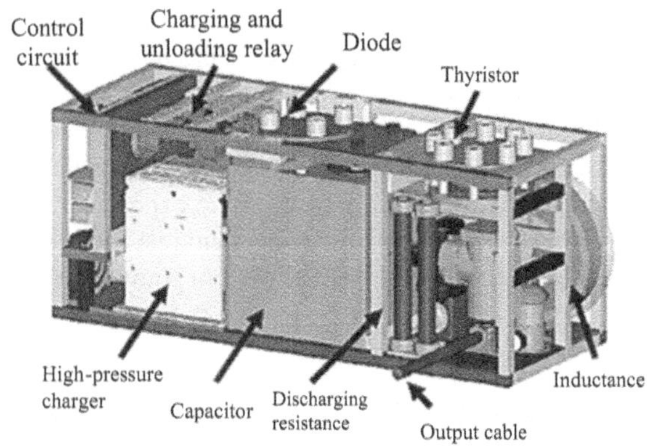

Fig. 7.10 3D model of 100 kJ pulsed power supply system for 120 mm self-electrothermal chemical gun in the United States

7.3.3.1 Miniaturization of Power Supplies

From the point of view of future practical application, in order to meet the requirements of system integration, the pulsed power supply required by electrothermal chemical gun is developed in the direction of high density, miniaturization and high reliability. In the ETC with medium and large calibers mainly acts as a ignition power supply, whose energy storage ranges from about 100–200 kJ. it can be seen that charging and discharging are integrated with each other, in order to meet the requirements of the speed to be more than 10 rounds/min and the service life to be more than 1000 rounds (Fig. 7.10).

7.3.3.2 Research on Ignition Technology

In the continuous research and development of new type of propellant, electrode materials and forms, the study of plasma ignition mechanism and process, and different ignition methods, through the precision control of the ignition time and temperature compensation effect, the performance of electrothermal chemical guns can be improved. There are two types of plasma igniters currently being explored: Flash Plate Large Area Emitter (FLARE) and Triaxial Compression Plasma Igniter (TCPI), as shown in Fig. 7.11. In addition, the research on how to use external power for ETC ignition is on the way, only to trigger a plasma igniter with a small explosive force.

7.3.3.3 New High-Energy, High-Density Emission and Loading Technology

The ETC plasma ignition device can be filled with propellant with higher density. At present, the propellant that general ignition devices use is of a density of 1.1 g/cm^3. Under an electrical energy of 75 to 250 kJ, the filling density of ETC plasma ignition

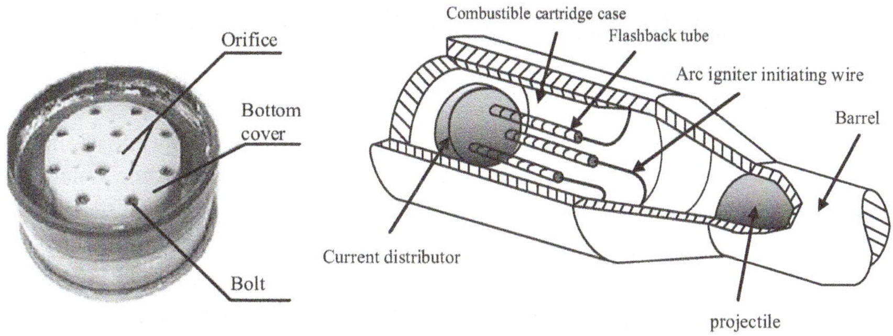

Fig. 7.11 The bottom-spray and triaxial compression plasma igniter of 120 mm electrothermal chemical guns

devices is more than 1.2 g/cm^3. The rising filling density of the propellant will increase the pressure at the bottom of the projectile, thereby improve the kinetic muzzle energy. The main technical challenge of electrothermal chemical propulsion technology is to develop a new type of high-energy low-vulnerability propellant, that is, the electrothermal ETI technology of LOVA is still one of the development directions.

7.3.3.4 System Integration Technology

With the breakthrough and gradual maturity of the key technology of electrothermal emission, the equipment and use of electrothermal chemical guns have been included in the "agenda" of developed countries, and the focus of its research begins to shift to system-level technology. It's significant to pay attention to system integration technology, as well as system performance under the conditions of the real-world in the future, such as practicality, reliability, suitability, safety and electromagnetic compatibility. Figure 7.12 is a conceptual layout of the overall structure of the wheeled 120 mm electrothermal chemical gun in the U.S.

Fig. 7.12 The conceptual layout of the overall structure of the wheeled 120 mm electrothermal chemical gun in the U.S

7.4 Joint Simulation Model of the Core CPA and Its Load System

Compared to electrothermal chemical guns, in which electrical energy and chemical energy work together, electromagnetic railguns only use the ampere force generated by current to advance projectiles, and their load characteristics are easier to express through theoretical analysis, so this section takes the electromagnetic railgun as an example and is combined with the finite element model of the air core CPA. Based on the finite element software Maxwell and the circuit software *Simplorer*, a joint simulation model of the air core CPA and its load system is established.

Under the Simplorer simulation platform, a two-phase air core CPA drive railgun system is established as Fig. 7.13, and the simulation model consists of three parts, namely Maxwell finite element model of pulsed alternator, switching device and controlling circuit of the system, and railgun model.

Taking two-phase dual-axis compensation air core CPA as an example, CPA's finite element model has four electrical ports, corresponding to the two-phase armature winding, compensation winding and excitation winding of the motor, and a mechanical port respectively. The inertia of the motor and the initial speed are given

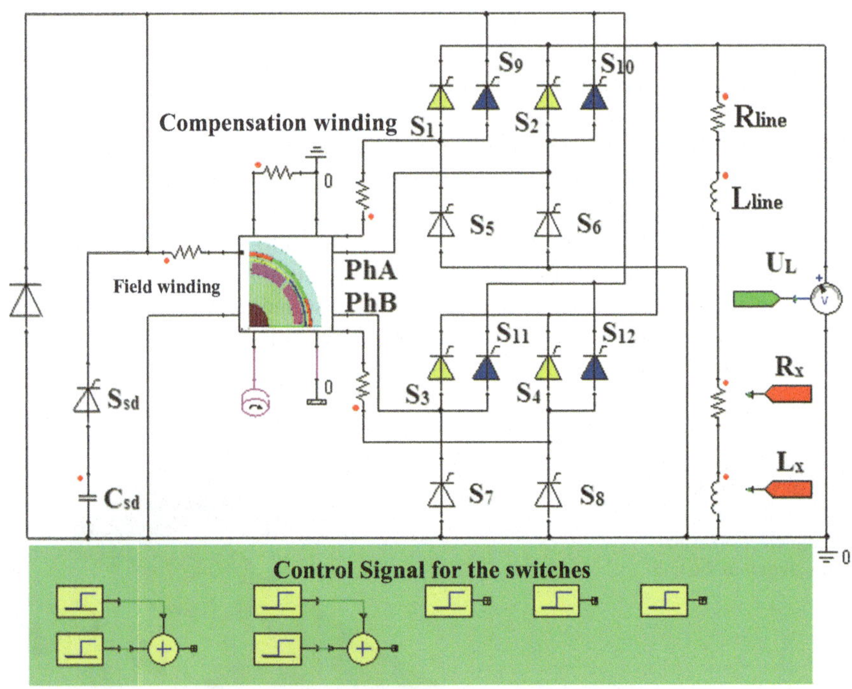

Fig. 7.13 Joint simulation model of pulsed alternator-driven railgun system

to describe the inertial energy storage of the motor and the potential loss in the running process.

In the external circuit, the reinforcement capacitor C_{sd} provides seed current, the switches S_1-S_4 are the upper bridge arm of the two sets of discharge rectifiers, S_9-S_{12} are the upper bridge arm of the two sets of self-excited rectifiers, and S_5-S_8 are the lower bridge arm shared by the two sets of discharges and the self-excited rectifiers, influenced by the respective trigger signal respectively.

The railgun model is a series circuit of variable resistors and inductors, each of which writes a value to each of them to simulate changes in parameters. Joint simulation calculation takes step as the smallest unit. In each step, after the calculation of the motor finite element and external circuit, the voltage U_L and input current i_L at the end of the railgun under this step can be obtained, and then the displacement of projectiles in the railgun can be calculated through the railgun subsystem. In this turn, the railgun's equivalent resistance and inductor under the step can be concluded. Finally, these parameters are written into the railgun model and moved on to the next step of calculation.

Chapter 8
Pulsed Power Switch Components

8.1 Classification

8.1.1 High Power Switch

Switching element occupies a special position in pulsed power system because the parameters and characteristics of switching element have the most direct and sensitive influence on the rising time, amplitude, etc. of pulse. The switch works under high voltage and high current, which means the working conditions are not very ideal. The breakdown in the switch and the discharging process in the switch electrodes are both very complex. Therefore, the pulsed power supply puts forward high requirements for the switch. The inherent inductivity and resistance of the switch are expected to be very small. The recanalizing delay should be short, and the dispersion should be small. It's also supposed to have favorable dielectric recovery and high repetition frequency, along with high voltage, high current, and long life-expectancy. A compact structure with small volume and light weight is also necessary. At present, the main switches used in pulsed power modules domestically and abroad are: Triggered Vacuum Switch—TVS, spark switch, RSD and thyristors. The main features of the various switches are briefly described as below.

(1) Trigger vacuum switch

Trigger Vacuum Switch, (abbreviated as TVS in the following context), have various advantages, such as accurate triggering, high voltage tolerance, wide operating voltage range, rapid recovery of main gap dielectric strength. It's also known for its short delay of action and dispersion, and its immunity to the external environment. However, its substantial size, instability, huge and complex trigger circuit limited its capabilities to fulfill the requirements proposed by the future high-power pulsed power supply, such as compact installation and flexibility for mobile use.

S. Wu and S. Cui, *Pulsed Alternators Technologies and Application*,
https://doi.org/10.1007/978-981-33-4224-8_8

(2) Spark switch

Spark switches have the advantages of high surge absorption capacity and working frequency. However, its tolerance voltage is quite low. It's also rather challenging to adjust its working point to the most ideal position on the Paschen curve, therefore it's also not practical to let the switch be discharged in the form of a "spark". In this case, the spark switch can be considered for circumstances in which high operating current, high pulsed repetition frequency and low operating voltage requirements are observed. In 2007 Germany already used spark switches in the PFN, with a forward blocking voltage of 70 kV and a peak current of 6 kA.

(3) RSD

In the early 1980s, I.V. Grekhov proposed a new power semiconductor device similar to a thyristor, as in the Reversely Switching Dynistor, abbreviated as RSD. Compared to traditional semiconductor switching devices, it has strong surge absorption capability, high di/dt, long life expectancy, low cost, and small loss. It's easy to be connected in series or paralleled, so a wide range of applications can be expected. But now the domestic researches are still on a low level. The developed RSD cannot withstand reverse voltage, and the trigger circuits and the magnetic switches in series connections are large. Thus, it's difficult to apply them into pulsed power supply.

(4) Thyristor

In the practical application, the thyristor appears to have high efficiency, good control characteristics, long life expectancy, small size and strong function, and its voltage and current capacity is the highest among the current power electronics, with high reliability. Because of the above advantages, a lot of work has been done on its application in the field of pulsed power sources abroad, and it has been adopted by many pulsed power energy modules as the main switch. The domestic high-power thyristor is mainly used in industrial frequency environments with high-voltage DC transmission. However, due to the high voltage, high current, and high steepness of the pulsed power supply, the technical parameters within the industrial frequency environment is not enough to reflect its use in the pulsed power supply accurately. There are barely applications in the field of pulsed power supply research. It is still waiting for more experiments and explorations. Based on this application background, this paper provides theoretical simulation and experimental research on the operation and damage factor under high voltage and high current conditions, as well as the protection when used in pulsed power supply system. This paper aims to explain the correct use of thyristors in pulsed power supply system on the basis of theoretical and practical foundations.

8.1.2 *Application of Pulsed Thyristors in Electromagnetic Emission*

Among high-power semiconductor switching devices, thyristors are currently the devices with the highest voltage resistance and greatest current capacity. The main production of high-power thyristor at home and abroad are used in high-voltage DC transmission. For the current high-power thyristor, the maximum diameter can be 6 inches, and the voltage resistance of a single valve chip can be up to 11 kV. Under operating conditions, a flow capacity of no more than up to 4500 A can be observed. Leading in this field are ABB of Switzerland and Zhuzhou Nan Che Shi Dai Co., Ltd. of China.

In order to improve the flow capacity, opening speed and *di*/*dt* tolerance of the thyristors, two new types of thyristors have been developed abroad on the basis of the ordinary ones. They are Gate-turn-off Thyristor (GTO) and Integrated Gate Commutated Thyristor (IGCT). Both devices have been put into practical use abroad. Specifically, GTO's single-chip voltage resistance can be up to 4.5 kV. Under operating conditions its flow capacity can be 4 kA for maximum. While the maximum voltage of the current IGCT in power systems achieves 10 kV and its flow capacity reaches 6.5 kA.

For thyristors applied in the pulsed power supplies, there are no domestic manufacturers who carried out any research in this field. Swiss ABB company still champions worldwide. It has been studying on high-power thyristors for pulsed power supply for more than ten years. The more mature devices currently used are GTOs with a diameter of 3.5 inches and a single chip of voltage resistance of 4.5 kV, usually working in series with three valves. The maximum current and the *di*/*dt* of current rise rate it can withstand is 120 kA/90 μs and 3.5 kA/μs respectively. The gate electrode tolerates a maximum trigger current of 800A and a *di*/*dt* of triggering current rise rate of 400 A/μs. Its newly developed IGCT has better performance, with a diameter of 3.5 inches and a single valve chip voltage resistance of 4.5 kV. The maximum current flow capacity is already up to 180 kA/30 μs and the maximum sustainable current rise rate *di*/*dt* is 20 kA/μs. The biggest triggering current that the door can cope with is 2000 A, and the maximum trigger current rise rate *di*/*dt* is 1000 A/μs. However, the reverse voltage that such switches can deal with is low and therefore can only be used in specific pulsed power supplies.

8.2 How Pulsed Thyristors Work

Thyristor, also known as Silicon Controlled Rectifier (SCR), was invented by Bell Laboratories in the United States in 1956. In 1957, General Electric Company developed the world's first thyristor and commercialized it in 1958. At present, the applications of thyristors in many fields are already very mature technologies, such as high voltage DC transmission. In the field of pulsed power technology, because the voltage level and current capacity of the thyristor are the highest of all high-power semiconductor devices. It thus becomes the preferred device in the field of high-voltage pulse. The thyristors are used under the condition of single large surges, and there is no relevant experience domestically. To fill this gap, this paper mainly studies on the electrical characteristics of thyristors applied in pulsed power supplies.

8.2.1 Basic Structures and Working Principles

The working principles of the thyristors can be explained by the dual transistor model, as shown in Fig. 8.1a. The interior of the thyristor is a four-layer semiconductor structure, named P1, N1, P2 and N2 respectively. It consists of three PN knots, J1, J2, and J3. There are three induced electrodes externally: the anode A from the zone P1; the machine G from the zone P2; and the cathode K from the region N2. By taking an oblique cross-section on the device, the thyristor can be treated as a combination of two transistors, VT1 and VT2, consisting of P1N1P2 and N1P2N2, as shown in Fig. 8.1b. The total emitter current amplification of VT1 and VT2 are $\beta_1$and $\beta_2$respectively. Both transistors are amplified when the anode is added to the positive voltage relative to the cathode. At this point, if a positive trigger signal is applied to the gate electrode, i.e. when there is a gate current I_g flowing in, there will be a base current flowing through VT2. Amplified by VT2, the collector current is $I_{e2} = \beta_1 I_{b2}$. Because the collector of VT2 is directly connected to the base of VT1, the base current of VT1 can be expressed as $I_{e1} = I_{c2}$. The current I_{c2} is amplified by VT1, whose collector current $I_{c1} = \beta_1 I_{b1} = \beta_1 \beta_2 I_{e2}$ is hereby obtained. When I_{c1} flows back to the base of VT2, a positive feedback is formed, which makes I_{e2} increase continuously. The two triplets soon enter the saturation state, that is, the trigger-on state of the thyristors.

According to the working principles of thyristors, the following equations can be put forward as

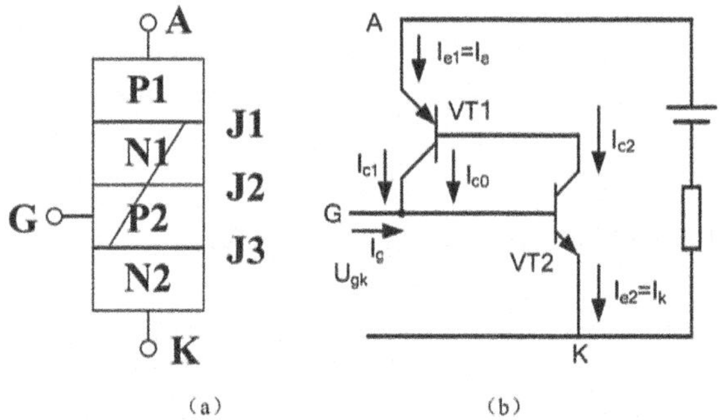

Fig. 8.1 Thyristor topology

$$\begin{cases} I_{c_1} = \alpha_1 I_a + I_{CBO_1} \\ I_{c_2} = \alpha_2 I_k + I_{CBO_2} \\ I_k = I_a + I_g \\ I_a = I_{c_1} + I_{c_2} \end{cases} \tag{8.1}$$

where $\alpha_1$and α_2 are the co-base current gain of transistors VT1 and VT2, respectively. I_{CBO1} and I_{CBO2} are the co-base extreme leakage currents of VT1 and VT2, respectively. From the above equations, it can be deducted that

$$I_a = \frac{\alpha_2 I_g + I_{CBO_1} + I_{CBO_2}}{1 - (\alpha_1 + \alpha_2)} \tag{8.2}$$

The characteristic of the transistor is that it can be very small at a low emitter current α. When the emitter current has been built up, α it increases sequentially. Therefore, when the transistor is in a blocking state, there are $I_g = 0$. However, $\alpha_1 + \alpha_2$ is very small. As can be seen from above, the leakage current running through the thyristor at this moment is only slightly greater than the total current leakage of the two transistors. If the injection of triggering current increases the current in each transistor's emitter and get $\alpha_1 + \alpha_2$ close to 1.0, the current in the thyristor I_a will approach infinity, thus achieving the saturation of the device. Of course, due to the limitation of the external circuit load, I_a will actually remain limited. At this point, the gate voltage is cleared, and the thyristor remains on.

8.2.2 The Working Characteristics of the Pulsed Thyristors

Semiconductor components that are able to handle surge current is also one of the parameters of its on-state characteristics. Usually the rated working current of high-power thyristors work on the kA level, especially under the condition of power frequency or DC. It's still rarely seen domestically to be used in the field of pulsed power technology. In Pulsed Forming Networks, the current passing through the thyristor rises to a quite high current peak (10–100 kA) in a very short period of time (on microsecond level). The surge current ITSM that a thyristor is able to bear is usually 10 to 20 times its normal operating current. The two main factors limiting the pulsed current tolerance of the thyristor are tolerances of the peak current and the *di/dt* of the current rise rate.

(1) Flow capability

The ability of the thyristors to hold pulsed current depends on $\int i(t)dt$. The pulse width of the pulsed current is very narrow relative to the power frequency or DC current. Therefore, under the condition of a single pulse, the thyristor can cope with a decent amount of pulsed current peak value. Usually there are two reasons for the pulse current peak to damage the thyristor.

The first one is electric breakdown. When the pulsed current goes through the thyristor, an overheat from overload took place, which reduces the reverse blocking capacity of the thyristor. After the pulsed current passes through, the component is knocked out in reverse when a higher reverse voltage occurs in the thyristor.

The second one is thermal breakdown. This arises from local concentration of heat. Because the thyristor is small and light, its package is vacuum sealed, its heating is particularly concentrated. When the pulse current passes through, the valve chip current density is high. The greater the current is, the higher the temperature of the hot spot is, the smaller its body resistance will be. In this regard, the current will be greater, the temperature will further increase, so that the circulation continues to develop. This will make the current shrink to the hottest part of the channel in a short period of time, causing the local temperature exceeds the critical temperature, resulting in a disastrous burnout.

Based on the above introduction, it can be seen that the current peak tolerance of the thyristor has a great deal to do with its working temperature. Therefore, in order to improve the thyristor's ability to withstand current spikes, adjustments can be made to the component's manufacturing process or internal structural parameters, such as maintaining sufficient residual life of the minority carrier in the base area and increasing the diameter of the wafer. It's also necessary to ensure that when the thyristor is subjected to pulsed current, the working temperature does not exceed the allowable value. Its flow capacity is also expected to be improved.

(2) Limits on the rate of current rise

When the thyristor is on, if the current flowing through is very large or the rise time is very short, it will increase the *di/dt* of the flow passage. Since the expansion process of the carrier takes some time, most of the current is concentrated in the initial triggering on-state region. The current density in this region will be enormous. This may cause intense local heat and may make the device work improperly or even burn. Thus, for the thyristor, the current rise rate, that is, commonly referred to as *di/dt* tolerance, can be a real problem. The *di/dt* tolerance is usually related to the size of the initial opening area of the thyristor and the diffusion speed of the carrier. High *di/dt* causes damage to the thyristor mainly due to thermal fatigue damage. Because of the triggered flow of the triggering electrode, there appears dense current near to local firing points. This leads to a strong heating and elevated temperature of some components. Thermal stress caused by temperature mutations can give rise to lattice damage to the thyristor materials, which eventually causes complete damage to the SLR. Such mechanism is known as thermal fatigue damage. In this case, the temperature of the hot spot becomes very high, so that a longer shut-down time takes place. This reduces the thyristor *di/dt* tolerance and deprives its abilities to work. In order to increase its *di/dt* capacity, it is necessary to increase its on-expansion speed. Less opening time of the thyristor may help or improving the trigger electrode-cathode structure to increase its initial on-area, or to increase the distance between the gate electrode and the cathode, so that the trigger power will be enhanced.

8.3 Series Thyristor Assembly Protection

When a switch with high tolerance to voltage is required, the voltage resistance of a single thyristor is limited, and a single thyristor cannot meet the voltage-resistant demand. It is necessary to connect multiple thyristor in series to meet the conditions of the switch. In the applications of the device, due to the static Volt-ampere characteristics and dynamic parameters of each component, it will cause uneven voltage distribution between the components and lead to a damage to the devices. There are several main influential factors for the uneven distribution of voltage in series operation (Fig. 8.2):

1. The effect of static Volt-ampere characteristics on static average voltage. The Volt-ampere characteristics of different components vary widely, and the voltage distribution is uneven when used in series. At the same time, the Volt-ampere characteristics of semiconductor devices are susceptible to temperature. Different joint temperatures can also impose some impacts on voltage equalizing.
2. The effect of dynamic characteristics such as turn-off charge and opening time on dynamic voltage equalizing. The thyristors run in series with different delays. The size varies from different gate triggering pulses. These will give the valve chips different opening speeds. Different opening speeds of valve chips give rise to uneven dynamic voltages. Meanwhile, the differences in shutdown time will also turn the thyristors off at different times. If there are few turn-off charges, it will be easy to shut down, and the shutdown time will also be short. The component which was shut off initially bears the highest dynamic voltage.

The fundamental purpose of thyristor series technology is to ensure that the thyristors with different dynamic and static characteristics can operate safely and steadily and be fully utilized afterwards. This involves a series of problems such as component protection, dynamic and static equalized voltage trigger consistency, reverse recovery of over-voltage suppression, and buffer on and off.

There are five states in the whole on and off process of the thyristor, namely, positive blocking, opening process, on state, ability recovery blocking, and reverse

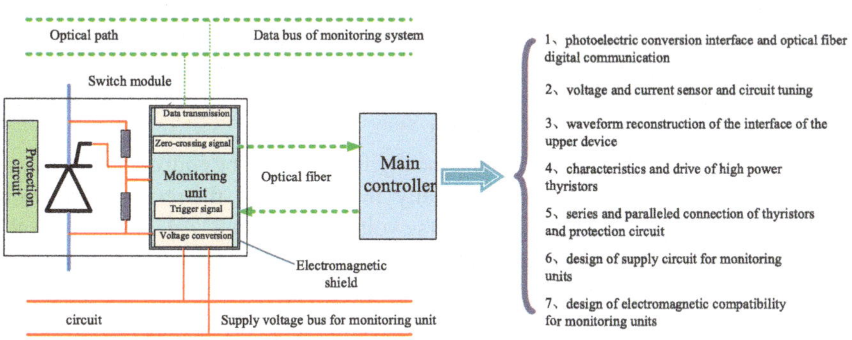

Fig. 8.2 System control block diagram

blocking. In the thyristor series, the different on-state characteristics have little effect on the split voltage, while the other states have different degrees of effects. The equalized voltage of the positive and reverse blocking is called static voltage balancing, and the equalized voltage of opening and blocking the recovery process is called dynamic voltage balancing.

8.3.1 The Basic Principles of Unbalancing Thyristor Series Voltages

When the thyristor is in the positive blocking state or reverse blocking state, all series of thyristors are subject to a steady voltage drawdown. Its distribution is determined by the Volt-ampere characteristics of each thyristor. At this point the equalized voltage problem between them can be categorized into static voltage balancing.

In the positive blocking state, the thyristor withstands a positive steady voltage drawdown. Due to the controlled conductivity of the semiconductor device, the thyristor is not tolerant to a large current, but a very small leakage current is exceptional. The voltage distribution on the series of thyristors is determined by the positive Volt-ampere characteristic of the device. Under the same leakage current, the thyristor with different positive Volt-ampere characteristics will have different voltage distributions, and the difference in the leakage resistance of the positive blocking area of the individual thyristors leads to the unbalancing voltage in the positive blocking transient.

There is a similar steady balancing voltage problem in the reverse blocking state. Because of the discreteness of the thyristor parameters, the sizes of the reverse leakage currents in each thyristor is difficult to be completely unified, so there will be an imbalancing voltage distribution in the reverse blocking state.

Due to the differences in thyristor parameters such as condensing capacitor and carriers, different thyristors are used differently in the positive opening process or reverse blocking recovery process. This results in the problem of unbalancing voltage in these dynamic processes.

During the positive opening process, the anode voltage and anode current of the thyristors are in a dynamic process, in which the thyristor can be equivalent to a time-varying resistance that changes exponentially. If the opening time is not consistent, a dynamically unbalancing voltage in the opening process will be derived and the devices opened later will be overwhelmed.

During the reverse blocking recovery, the thyristor can still be equivalent to a time-varying resistance that changes exponentially. In this process, the reverse current of the thyristors gradually decreased. A reverse voltage began to be established at both ends of the device. If the reverse recovery charge of each series is not consistent, then the thyristors which is firstly recovered to reverse blocking will come across overvoltage, so that there will be a problem of unbalancing voltage on the series of thyristors.

8.3.2 Static Voltage Balancing Design

In order to meet the requirements of the design of voltage balancing, first of all, the favorable choices for devices are those with consistent parameters and characteristics. Then, at both ends of each thyristor, a paralleled resistance branch needs to be implemented for static voltage balancing, and a parallelled RC branch for dynamic voltage balancing. These are to reduce the difference between the various devices in terms of turn-on delay.

The main way to solve the problem of static voltage is to parallel the thyristor with an equalizing resistance, with whose voltage to limit the differences between the blocking voltages of the series thyristor to a certain range. In the states of positive blocking and reverse blocking, the rising edge at the end of the thyristor is rather wide, time-consuming, and the equivalent resistance is high. In this circumstance, it's possible to carry out paralleled resistance on both sides of the thyristor for the sake of shunts. It's given that the value of the paralleled resistance is approximately equal to the minimum value of each resistance within. Thus, after paralleling each thyristor with resistances of equal values, the resistance of each shunt unit becomes basically equal. Then there the distribution of voltage in the shunt unit will be more balancing. Here comes a favorable static voltage balancing effect.

8.3.3 Dynamic Voltage Balancing Design

Dynamic voltage balancing comes from the inconsistency of the turn-on characteristics of the thyristors. An effective option to solve this problem can be a paralleled capacitor resistance branch. This is based on the principle that there will be no mutation of voltages at both ends of the capacitor. The capacitors of the same tolerance are paralleled at both ends of each series thyristor to absorb momentary strong currents and decelerate the voltage changing rate on the component. The role of the accessed resistor is to prevent the capacitor from oscillating with the stray inductance in the circuit, and to prevent the thyristor from discharging to the device sharply when the capacitor is on, and to suppress *di/dt* of the current rise rate.

8.4 Effect of Gate Electrode Structure and Triggering Circuit on Thyristor Turn-On Characteristics

From the perspectives of the influence which is imposed on the turn-on time from inherent structure and process of the thyristor, the length of its delay depends on the transit time of the internal carrier of the valve, and the size of the rising time depends on the accumulation time and diffusion speed of the internal load of the valve. Therefore, the gate electrode structure and process of the thyristor is very influential to its turn-on time. From the perspective of application, the length of turn-on period is closely related to the characteristics of gate currents, such as its strength, as well as the wave front steepness and the width of the gate pulse. Turn-on time is associated not only with the gate triggering characteristics, but also with the anode voltage applied to both ends of the SCR element and the anode current passing through the SCR element. Increased anode voltage reduces turn-on time. Increased anode current has little effect on the length of delay, but not for the rising time. The rising time is lengthened accordingly.

8.4.1 Effect of the Trigger Electrode Shape on the Opening of the Thyristor

The thyristor gate electrode structure also has a great influence on its turn-on characteristics. The change of the gate electrode structure has a great effect on shortening the turn-on time of the thyristor and improving its *di/dt* capacity. The usual method is to extend or enlarge the gate electrode. Both of these can increase the initial turn-on area and speed up the thyristor.

8.4.2 Effect of the Trigger Pulse on the Turn-On State of the Thyristor

The function of the thyristor trigger circuit is to produce a gate trigger pulse that meets the requirements, so that the thyristor can be turned on normally when needed. The thyristor trigger circuit is expected to fulfill the following requirements:

(1) The trigger pulse should be wide enough to make sure that the thyristor can be turned on.
(2) The trigger pulse should have sufficient amplitude. For some occasions with low temperature, the magnitude of the pulse current should be increased to 3 to 5 times the maximum trigger current of the device. The steepness of the pulse also needs to be increased;
(3) The trigger pulse provided should not exceed the voltage, current and powering rate of the gate electrode of the thyristor. It should be within the reliable trigger region of under the Volt-ampere characteristics of the gate;
(4) The trigger should have good anti-interference ability, temperature stability and electrical isolation from the main circuit;
(5) The trigger pulse should help the thyristor elements' turn-on times to be consistent. In the thyristor series circuit with high voltage and current, the components in series are required to be turned on at the same time. In this case, a strong trigger should be used.

When the pulse width of the trigger current remains consistent, the current peak and rising time of the trigger current have a great influence on the turn-on state of the thyristor. Both mentioned above can improve the breakdown energy of the gate PN junction of the thyristor and shorten the delay. The delay-time curve of different SCR elements is greatly dispersed. However, the general trend is that with the increase of the trigger current, the delay time of the thyristor is gradually shortened. As the rise time of the trigger increases, the delay time of the thyristor also increases significantly. It thus can be seen that the rising steepness of the trigger pulse also has an effect on the delay time of the thyristor.

At the same time, increasing the threshold trigger amplitude and shortening the rise time of the gate trigger, that is, increasing the forward steepness of the pulse, is not only beneficial to reduce the delay time, but also has a more obvious effect on shortening the rise time in the opening time. The rising amplitude and steepness of the gate trigger increased the initial turn-on area of the SCR element. Hereby the expansion speed of the on-zone is accelerated, the expansion time is shortened, and the turn-on of the SCR element is accelerated. A shortened rising time is thus achieved.

Therefore, in order to improve the performance of a turn-on thyristor, in the actual application of the trigger circuit design, the current peak of the trigger current will be escalated as much as possible, and the rising time will be reduced as much as possible. However, an overly strong pulse will cause a transient increase in gate power, which may exceed the limit peak power of the gate that the thyristor can withstand. This will over-heat the gate until it burns out. Therefore, there should be an appropriate limited range for the peak of the output waveform of the trigger circuit.

8.5 Simultaneous Trigger of the Pulsed Thyristor Assembly

The trigger circuit of series thyristor assembly is required to have the following characteristics:

(1) The trigger circuit must have reliable insulation ability to ensure the safety of the control circuit on the low-voltage side;
(2) In order to prevent an over-voltage breakdown, when triggering the series thyristor, the steeper the front of the gate trigger pulse is, the more likely it will be for the series thyristor to be consistent.

8.5.1 Electromagnetic Trigger

Electromagnetic trigger circuit uses pulse transformer to trigger the thyristor tube and realize the isolation between high and low voltage circuits. Advantages of high reliability, simple circuit and low cost are implied.

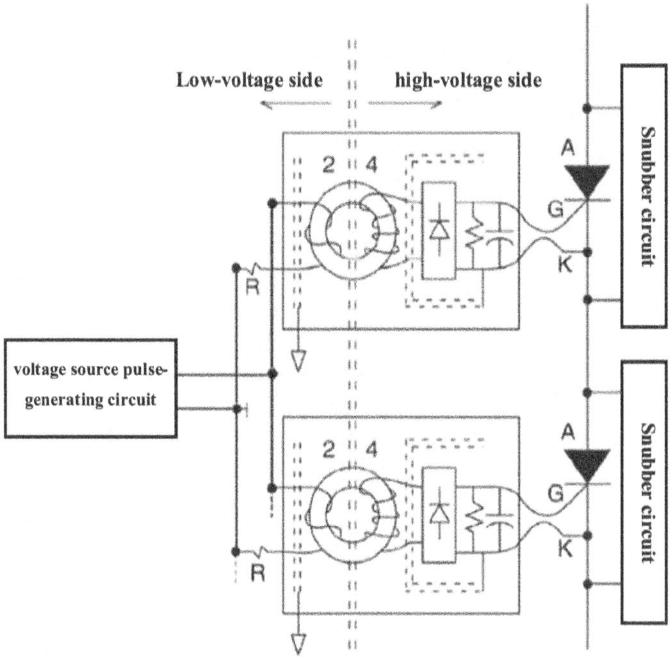

Fig. 8.3 The voltage source drives the pulse transformer

(1) The voltage source drives the pulse transformer (Fig. 8.3)

Each thyristor in the circuit is triggered by a pulse transformer, which is installed near to the thyristor. All pulse transformers are paralleled. A voltage source pulse-generating circuit is used to drive these pulse transformers. Depending on the voltage applied by the low-voltage side of the pulse transformer, a pulsed transformer with a suitable transformation ratio can trigger the thyristor.

This electromagnetic trigger circuit shares the characteristics of simple trigger circuit. However, the pulse transformers used to isolate the high and low voltage side make the pulse transformer large and bulky in size thus it's not easy to install.

(2) The voltage source drives the pulse transformer (Fig. 8.4)

Each thyristor in this circuit is triggered by one pulse transformer, which is installed near the thyristor. All pulse transformers are connected in series via a high-voltage cable that serves as a winding for these pulse transformers and insulates the high and low-voltage side. When using a high-frequency AC current source to drive these stringed pulse transformers, to trigger the series thyristor valve set, it's necessary to choose a suitable transformation ratio according to the strength of the high-frequency AC current source current. The pulsed transformer works in the same way as the current transformer when the trigger circuit is working.

Fig. 8.4 The current source drives the pulse transformer

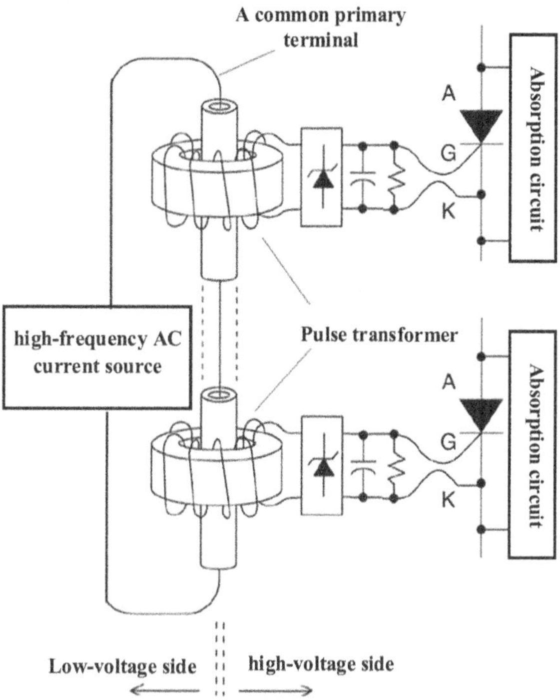

Such electromagnetic trigger circuit enjoys simple structure and small volume of pulse transformer, but a high-frequency AC current source for the driving circuit at the low-voltage side is required. The design of the very circuit is complicated.

8.5.2 *Direct-Light Trigger*

The direct-light trigger is to convert the trigger signal into a light pulse, directly triggering the high-level light-controlled thyristor. In the direct-light triggering system, the SCR element can be directly triggered by a light pulse with certain wavelength and certain energy intensity. After being changed into a light pulse, the point pulse directly acts on the SCR element gate electrode through the fiber transmission. This saves the high-level trigger circuit and the conversion circuit of the wide electrical signal and gives the trigger a higher reliability. Because there are basically no light-controlled thyristor products domestically, and foreign products are very expensive, it is not put this design into practice currently. However, the light-controlled thyristor can be a trend in the future (Fig. 8.5).

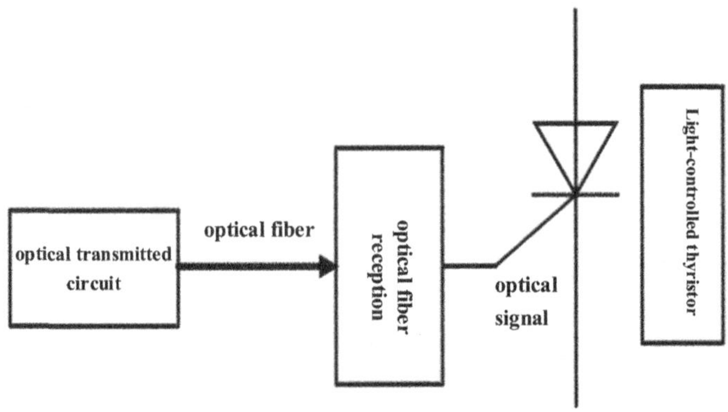

Fig. 8.5 Direct-light trigger

8.5.3 *Indirect-Light Trigger*

Indirect-light trigger is essentially optical fiber communication. The trigger signals are sent from low levels to high levels and are processed to couple with the thyristor gate electrode through one or more photoelectric conversions. Its working principles are shown in the figure above. The advantage of such trigger is that the trigger signals are not subjected to electromagnetic interference during transmission. This guarantees the strength and steepness consistency of the trigger waveforms. The cost is also easily to be cut due to its adoption of normal thyristors. In particular, in the application of pulsed power technology, where insulation and electromagnetic compatibility are crucial, the excellent insulation performance of optical fiber can reduce the cost of insulation and isolate the electromagnetic interference from the high-voltage side to the low-voltage side (Fig. 8.6).

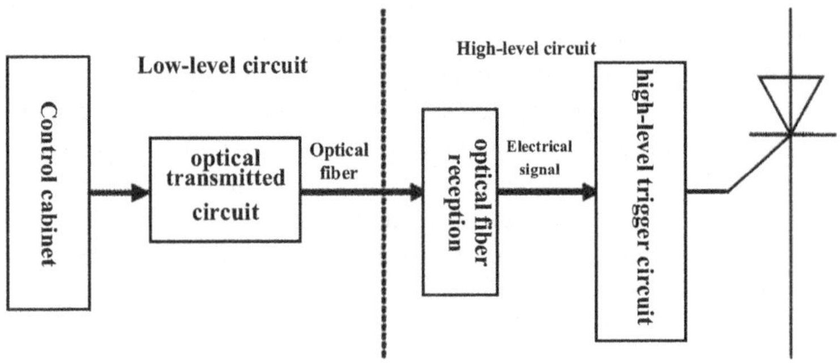

Fig. 8.6 Indirect-light trigger

Chapter 9
Pulsed Alternator Drive System

9.1 Introduction to the Pulsed Alternator Drive System

Drive system, as the energy input component of the pulsed alternator power supply, plays a decisive role in the performance of the whole system. During the initial energy storage process, the prime mover drives the pulsed alternator to speed up from 0 to the rated speed, this process converts the electrical energy into mechanical energy in the rotor of the pulsed alternator. The prime mover is then disconnected from the power supply system, and the pulsed alternator discharges. After discharging, the prime mover is re-connected to the power supply, driving the pulsed alternator to accumulate to the rated speed again. The pulsed alternator hereby gets the energy to recover, so it can discharge and then go back to the energy storage and discharging process. According to relevant research, several generations of pulsed alternators in the United States have experienced numerous changes. Their drive systems developed from hydraulic motors, transmissions, clutches to low-speed motors, transmissions, clutches, and high-speed motor driven directly by current. The recent actualized transformation of pulsed alternator power supply takes the miniaturization of power supply as its prerequisite. A complicated drive system is neither conducive to reduce the size and weight nor raise the reliability of the system. Therefore, it is very important to study on the direct drive of a high-speed motor for the sake of the whole pulsed alternator power supply system.

9.2 Development of High-Speed Motors

High-speed motor usually refers to the motor with a speed of more than 10,000 rpm. Claw-pole alternators and homopolar machine have become high-speed motors, but mainstream ones are DC motors, induction motors, switched reluctance motors and permanent magnetic synchronous motors.

S. Wu and S. Cui, *Pulsed Alternators Technologies and Application*,
https://doi.org/10.1007/978-981-33-4224-8_9

DC motor distinguishes itself by its simple control, good speed adjusting performance, wide speed range, and large starting and braking torque. It's also easy to achieve a quick start-up and parking. Therefore, it was popular for the early high-speed drive system such as NC machine tools. However, because of its structure is based on separate or series excitation, the stator is composed of the main magnetic pole and the communicating pole, while the rotor is composed of the armature winding and the commutator. These make the DC high-speed motor structure more complex and larger. In addition, its reliance on the electric brush and the commutator to change the current direction of the armature winding will produce a commutation spark. This will deteriorate the commutation and intensify the wear of the brush and the commutator, even lead to malfunction. According to relevant literature review, at the end of 1985, the output power developed in Japan was 220 kW, and the DC motor with a speed range of 2000 rpm to 8000 rpm was the highest technical level of DC motor overworld at that time. Although the processing level and performance of DC motor brush and sliding ring have been greatly improved over the years, the problems lie in brush and commutator are still quite fatal. They restrict the development of DC motor to higher speed and higher power. At present, the maximum speed of DC motor is about 20,000 rpm, and the speed of micro-DC motor can be 30,000–40,000 rpm.

Induction motor has small torque ripple and there is no mechanical commutation. The squirrel-cage rotor is with simple structure and advanced mechanical balance. Its good start-up performance and weak magnetic property are ideal for long-term operations under high temperature and high speed. Another advantage of induction motor is that they do not require position sensors and can work in open-loop states at a much lower price than DC motors of the same speed and power. Therefore, induction motors are widely used in high-speed applications. A high-power, high-voltage, high-speed induction motor (8 MW, 6.6 kV, 12,000 rpm) has been developed by scientific research institutions. It is applied in natural gas compression equipment for studies on the selection of the materials, end effects and its correction coefficients for the solid rotors. In addition, MIT's Electromagnetic and Electronic Systems Laboratory developed a 5 MW high-speed induction alternator. Texas State University Mechanical electronics center developed a high-speed induction fly-wheel motor for advanced locomotive propulsion systems. ABB developed the world's largest high-speed induction motor at the time in 2002, with a speed of 20,000 rpm and a power up to 15 MW. Based on past studies, the induction motor with a maximum speed of 180,000 rpm and a power of 10 kW high-speed has already been used for testing purposes. With five-axis active magnetic levitation bearings, and a linear velocity speed of 219 m/s, the air friction loss of the motor can be up to 690 W, which is equivalent to the corresponding iron loss. The efficiency of the motor in this case is about 85%. The huge shear stress in the high-speed operation and thermal expansion of the rotors motor lead to rotor deformation. This greatly increases the imbalance of the motor. It needs to be improved in practice even though the high-speed induction motor of magnetic bearing technology with a power of 1.6 kW and a speed of 180,000 rpm has been successfully applied to the actual equipment. However, high-speed induction motor have some disadvantages that can't be ignored: (1) the current base wave and harmonic components in the rotor circuit will produce a large rotor vortex loss; (2) In order to overcome the huge shear stress of the rotor at high speeds, metal materials with long axes and more strength are better choices, which will cause the rotor resistance to be considerably high. If the sectional area of the rotor conductor is enlarged to reduce the rotor resistance, the rotor magnetic field will be saturated, and the excitation current will be strengthened. (3) The power factor of high-speed induction motors will be rather small.

Switched Reluctance Motor (SRM) is popular for its simple structure, sturdiness and durability, low cost, high fault tolerance degree, outstanding operation capability under high temperature, large amount of control parameters, and flexible control methods, etc. Switch magneto-resistor motor rotor is of convex structure, which means there is no winding, permanent magnet and slip ring on it. Steel sheet works in a nonlinear high magnetic dense zone. Its silicon wafer utilization rate is high, compared to DC motors and even induction motors. It also harbours small size, light weight, strong structure, as well as simple and convenient maintenance. Thus, switching magneto-resisting motors are particularly suitable for high speed requirements, and can be used as high-speed and ultra-high-speed motors. Ku Hans et al. developed a high-speed switch reluctance motor with a speed of 30,000 rpm for motorized spindle, and successfully achieved sensor-less control. In 2000, France developed an ultra-high-speed reluctance motor with a rated power of 1 kW and composite shaft. Its maximum speed can be up to 200,000 rpm. Recent years witnessed lots of domestic research on high-speed switch reluctance motor. Fan Dong et al. from Nanjing University of Aeronautics and Astronautics designed a switch reluctance motor magnetic levitation bearing switch with a power of 7.5 kW and a rated speed of 20,000 rpm. They studied on many issues including the motor's electromagnetic design, structural strength, stability design, rotor stress design, thermal design, magnetic material selection and other key technical issues. Chen Haijin et al. from Southeast University laid out a universal digital controller for high-speed switch reluctance motors using digital integrated circuits. They limited the speed in a range from 60 rpm to 45,000 rpm. The special structure of the switch reluctance motors has many advantages as well as disadvantages: (1) The torque's noisy and large ripple; (2) Lower density and efficiency compared to high-speed permanent magnet synchronous motor; (3) Due to the high degree nonlinearity caused by the saturated magnetic circuit, eddy current and hysteresis effect, and the characteristics of nonlinear switching power supply, the internal electromagnetic relation of the motor can be very sophisticated, and it is difficult to establish the mathematical relation with the conventional motor, which is not conducive to control; (4) To ensure a proper operation, the information about rotors positions must be obtained, so it is generally necessary to install position sensors. This increases the complexity of the structure while reduces the reliability of the system. Although there are a number of programs with sensor-less control, further development is required.

The main magnetic field of permanent magnet synchronous motor is generated by permanent magnet. This gives the motor distinguished high efficiency, high starting torque, high power factor and large speed range. It has gradually replaced asynchronous motor and electro-excitation synchronous motor and become the new buzzword in the field of high efficiency driving and high speed. The winding of permanent magnet synchronous motor can be in the centralized winding structure. The end of the winding is very short, the axial length of the shaft is therefore reduced. This is beneficial for the rot to develop a higher critical speed. In addition, the adoption of high-performance permanent magnet materials sets a quite low bar for the air gap. The high-speed motors still boast favorable performances when the motor air gap is about 5 mm. This is especially suitable for high-speed operation. On the one hand, this allows the outer protective layer to protect permanent magnets and improve its mechanical strength. On the other hand, it is conducive for the shaft to rotate in a higher speed than the critical one, reduce rotor loss and unbalanced magnetic pull. With the continuous emergence of new permanent magnet materials and the rapid development of advanced control technology, permanent magnet synchronous motor has become a new direction for high-speed motor development. In 1994, Takahashi et al. developed a surface-mounted high-speed permanent magnet synchronous motor with a power of 5 kW and a speed of 150,000 rpm. The length of its effective air gap is 6 mm. This was more than 10 times the ordinary permanent magnetic synchronous motor. Its rotor magnetic pole was made of Nd-Fe-B which enjoys large remaining magnetic. Its rotor sheath was made of fiberglass which was known for its tensile strength and high resistivity. The motor was supplied by a current source inverter and owned an efficiency up to 90%. There were also some scholars who put forward a high-speed permanent magnet synchronous motor for storage system with a power of 7.6 kW, a speed of 50,000 rpm, and 4 pairs of polar fly wheels. Magnetic circuit method and 2D finite meta-method were used for the studies on its armature reaction and demagnetization. NelcoSystems manufactured a 250 kW, 50,000 rpm turbine vane wheel-driven high-speed alternator in 1995 that acted as an electric motor when the turbine was started up. The system was about 600 mm in length, with a diameter of about 250 mm. The motor was 300 mm long with a 250 mm long diameter. C. Zwyssig et al. from Swiss Federal Institute of Technology Zurich developed an ultra-high-speed permanent magnet motor and a corresponding driving control system with a power of 1 kW and a speed of up to 500,000 rpm in 2007. Domestic research on high-speed permanent magnet synchronous motor is also in full swing. Yu Li et al. from Southeast University launched a high-speed permanent magnetic synchronous motor with a rated power of 600 W, and a rated speed of 20,000 rpm. They also carried out substantial studies on the iron loss and the temperature rise of the high-speed motor, theoretically and experimentally. Professor Wang Fengxiang et al. from Shenyang University of Technology modeled a high-speed permanent magnet synchronous motor rated at 60,000 rpm with a rated speed of 60,000 rpm and carried out in-depth research on key issues such as the structure and vibration mode of the high-speed motor. Professor Shen Jianxin from Zhejiang University laid out a high-speed permanent magnet synchronous motor with a speed of up to 150,000 rpm and a power of 2.3 kW. Professor Li Weili et al. from Harbin University of Science and Technology have developed a high-speed permanent magnet synchronous motors for distributed gas turbines at 117 kW and 60,000 rpm.

After comparing the defects of high-speed DC motor, induction motor and switch reluctance motor, the high-speed permanent magnet synchronous motor appears to be more in line with the demand for high-power and high-power density of pulsed alternators.

With the advancement of pulsed power technology in a more practical and applicable direction, there are gradually more requirements of miniaturization and modularization for power supplies. Minimizing the volume and weight of the system is now a decisive factor in the transformation. Rotor of pulsed alternator with large inertia is like a large fly wheel to the prime mover in terms of its long outside diameter. In order to make full use of the radial space of the system, shorten the axial length, and improve system reliability, the future goal for pulsed alternator is to integrate the driving motor with a pulsed alternator rotor and to connect them to the same axis. That is, part of the rotor is used to drive energy storage, while the other part is used to discharge. Hereby forming a module with fixed capacity and certain numbers of modules that fits the workload. The system puts forward the demands of printed high-power and high-speed for driving motors. There should be studies on the design and key technologies for printed high-power high-speed permanent magnet synchronous motor.

9.3 Key Technologies Within High-Speed Permanent Magnet Synchronous Motors

The key technologies within high-speed permanent magnet motor include: (1) Structural design and performance optimization as well as the effect of motor structure on electromagnetic performance. (2) High-speed motor rotor strength analysis, mainly refers to the protection of permanent magnet, etc. (3) Motor loss and thermal analysis.

9.3.1 Structural Design and Optimization

Structural design and optimization are to improve the electromagnetic performance of the motor, suppress torque fluctuations and reduce losses by changing the shape and type of fixed rotor core, winding, permanent magnet, sheath, etc.

High-speed permanent magnet motor topology has a crucial impact on the electromagnetic performance of the motor. Gradinaru et al. carried out classifications and comparisons on several common forms of permanent magnet motor and described their advantages and disadvantages as follows: (a) Slotless stator winding structure doesn't do much consumption to the iron core. It also reduces the iron loss of the fixed rotor. However, it increases the use of permanent magnet and copper, whereas more space is needed to for the stator winding. (b) Distributed stator windings have very small torque fluctuations, vibrations and noise. Because of the increase of its length at the end, there appears rising copper usage and consumption. However, fractional-slot stator structure may solve this problem. (c) The axial disc motor has

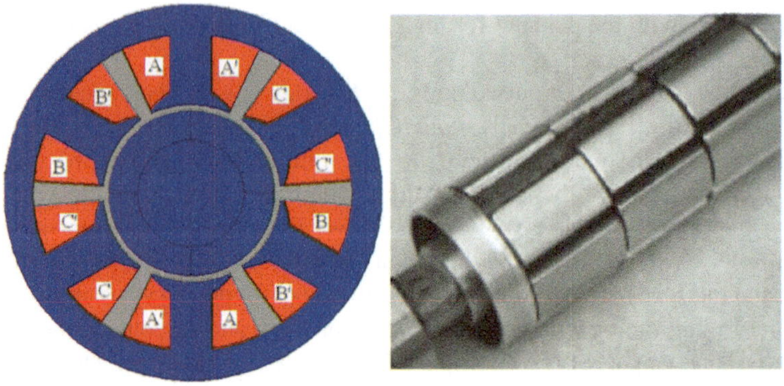

Fig. 9.1 A 25kW/15,000 rpm, 6/4 surface-type PMSM

a smaller rotor rotation inertia but requires more permanent magnets and is expensive to manufacture. They developed a 25 kW/15,000 rpm surface-type permanent magnet synchronous motor (PMSM) with 6/4 in-pole rotors. They used distributed windings to reduce harmonic weight, with the lowest cost and highest efficiency as the goal of optimizing the design of the motor. By changing the arc coefficient of the motor, magnetic shifting and skewed slot of the magnetic pole, the torque ripple of the motor can be reduced. The motor's sectional area is as shown in Fig. 9.1.

For traditional high-speed permanent magnet motors, their stators are slotted. Therefore, the optimization for these motors' performance can be achieved by adjusting the slot type, so that smaller air gaps can be obtained to reduce leakage of magnetism. Meanwhile, slotting increases the contact area between the stator winding and the fixed iron core, thereof enhance the thermal dissipation of the rotors and windings. However, the slotting of the fixed iron core will give rise to torque ripples of the motor, as well as more iron loss of the stator. Seok-Myeong Jang and other scholars from Chungnam National University in South Korea used analytical algorithm to analyze the cogging torque of high-speed permanent magnet motor. They pointed out that the slot of the stator opening could change the relative magnetic conductivity of the motor. They also suggested that this in turn would produce a spatial harmonic magnetic field, based on which the analytical calculation formula for the cogging torque of the permanent magnet motor was put forward. To extend the application of fly-wheel energy storage, they studied on the motor with slotless iron core structure of ring windings. They figured out that the ring windings would transfer the length at the end of the rotor yoke. For the motor with high power and few poles, this could effectively reduce the axial length of the motor. The structure without cogging slot can effectively eliminate the conventional problem of space harmonic magnetic field, and also reduce torque ripple, consumption and so on. The rotor permanent magnet within adopted Halbach structure, optimized the magnetic field waveform, and simulated its Back-EMF waveform which was close to the sine wave. The diagram of its model structure as shown in Fig. 9.2.

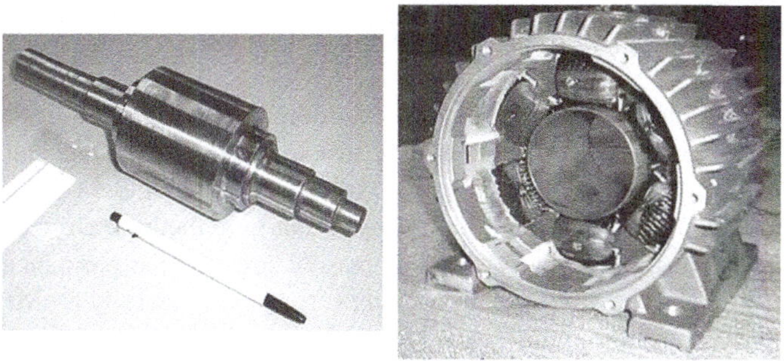

Fig. 9.2 A model structure of a high-speed motor for fly-wheel energy storage from Chungnam national university, South Korea

The structure of no cogging slot increases the air gap length. There are problems such as decreased utilization rate of permanent magnets and winding cooling difficulties etc. The current studies on high-speed motors only emphasize the effect of suppressing the cogging slots to the torques. While the effect of the method used on the electromagnetic torque of the motor is ignored. Thus, this is not the best for the electromagnetic performance of the motor. There are neither much research on the problems caused by cogging slot structures.

Larger permanent magnet bring greater sheath thickness problems. Confronted with the drop in the magnetic tightness of the air gap arose from the non-conductive magnetic sheath of the surface type permanent magnetic synchronous motor, J.M. Yon et al. put forward a semi-conductive magnetic laminated metal sheath as a compensation by increasing the thickness of the permanent magnet. Experiments confirmed that such method can improve the anti-electric potential by about 20%. The rotor structure after assembly as shown in Fig. 9.3.

Fig. 9.3 A diagram for the rotor structure with a semi-conductive magnetic laminated sheath

9.3.2 Rotor Structure and Strength

Surface-type permanent magnet synchronous motor is widely used in the field of high-speed and high-power due to its simple structure and high mechanical strength. Its permanent magnets are commonly made in cylindrical, cylinder or block or tile-type structure, using radial or parallel magnetic charging method, close to the rotor iron core. Considering that the yield compressive stress of the permanent magnet material is large and the yield tensile stress is very small (generally 80 MPa), in the case of high-speed rotation, the tensile stress effect of the huge centrifugal force on the permanent magnet is likely to exceed the ultimate tensile stress of the permanent magnet and cause damage. Therefore, the sheath pre-tightening force is applied outside the permanent magnet to protect it. There are two main ways of making sheath: one is to tie the permanent magnet with composite material (carbon fiber or fiberglass) and the other is to add a high-strength non-conductive alloy steel sheath to the outside of the permanent magnet. Regarding the different shapes of permanent magnets and the different types of sheath for the rotors, domestic and foreign scholars have done a lot of research. The following is a brief investigation and summary of the key references.

Andreas Binder et al. designed two 40 kW/40,000 rpm surface-mounted high-speed permanent magnet motors. One was with a pole arc coefficient of 0.87 and a very small non-conductive magnet at intervals. The simulation results showed that at the edge of the permanent magnet, the bandages would produce a lot of bending stress, and the sample rotor would break at high speed. The other's pole arc coefficient was 1. The permanent magnet was evenly distributed on the outside of the rotor iron core and added a layer of fiberglass inside the sheath. The bending stress from fiberglass was not significant. Such experimental results showed that the second motor could operate safely and reliably. The first sheath was installed under axial pressure, while the second one was installed by a direct sheathing with liquid nitrogen cooling rotors. The rotor structure of the two samples are as shown in Fig. 9.4, and the dimension data is shown in Table 9.1.

A. Borisavljevic analyzed the stresses on all parts of the rotor of a high-speed motor of 20,000 rpm. The permanent magnet of the motor was an injection-molded ring structure, part of which was buried in the rotor slot, as shown in Fig. 9.5. The size parameters of the motor are shown in Table 9.2. The protective sheath was made of carbon fiber material. When rotating at high speed, bending stress would be generated at the edge of the permanent magnet. A layer of fiberglass was first wound around the inside of the carbon fiber sheath in order to avoid carbon fiber breakage out of bending stress. In the axial direction of permanent magnets, fiberglass was used for fixed protection to prevent axial damage of permanent magnets. After that, the analysis model of the sheath material was established, isotropic and anisotropic. The stress in permanent magnets and sheaths were analyzed and calculated first and verified by Ansys finite meta-calculations. The experimental results showed that there was no big difference between isotropic and anisotropic models. These were basically the same with the results of Ansys finite meta-calculations. At the same

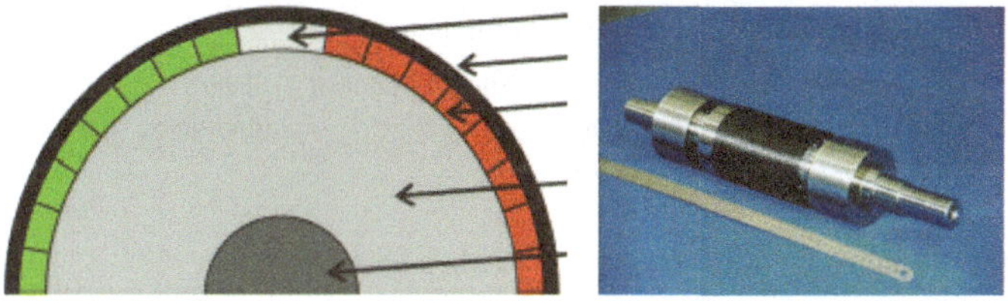

a) A sample rotor with a pole arc coefficient of 0.87

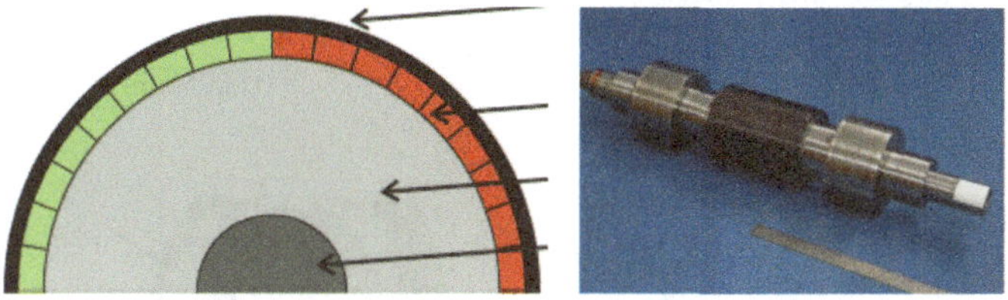

b) A sample rotor with a pole arc coefficient of 1

Fig. 9.4 Rotor structure diagram of two 40 kW/40 000 rpm high-speed permanent magnet motors

Table 9.1 Design parameters for two 40 kW/40 000 rpm high-speed permanent magnet motors

Pole arc coefficient	The thickness of the magnetic pole (mm)	Sheath thickness (mm)	Air gap (mm)	Sheath outer diameter (mm)	Magnitude of interference (mm)	Installation
0.87	7	5.7	0.7	88.6	0.24	Axis-side set
1	4.5	4.8	3.2	83.6	0.13	Rotor cooling

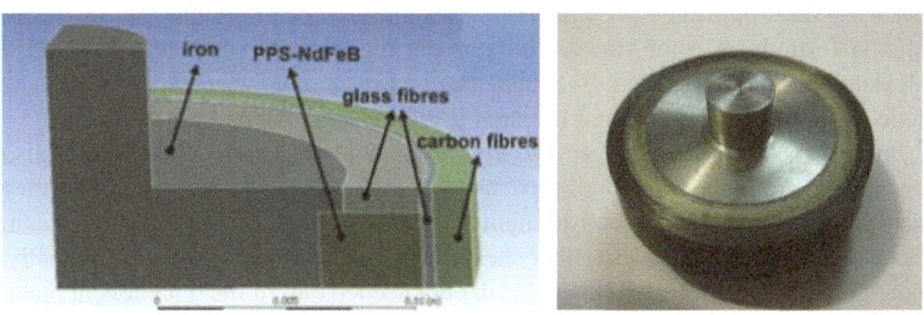

Fig. 9.5 Quarter sectional area of the rotor and its shape after transferring

time, the thickness of carbon fiber sheath was optimized and checked by analytic and finite meta methods.

At present, the domestic research on the strength of high-speed motor rotors are mainly from Shenyang University of Technology, Zhejiang University, Xi'an Jiaotong University and Shandong University of Science and Technology etc. Shenyang

Table 9.2 Dimension parameters of the 20,000 rpm motor

Speed	The inner diameter of the iron core slot	The outer diameter of the iron core slot	Sheath material	Sheath thickness	Glass fiber thickness
20,000 rpm	21 mm	23 mm	Carbon fiber	2 mm	0.5 mm
Permanent magnet thickness	Permanent magnet inner diameter	Permanent magnet axial length	Iron core axial length	Magnitude of interference	
4 mm	21 mm	6 mm	8 mm	0.0095 mm	

Fig. 9.6 Structure diagram of the 75 kW/60,000 rpm prototype

University of Technology started earlier than others. Professor Wang Fengxiang and others launched a design of the rotor structure of the metal sheath of a 75 kW/60,000 rpm high-speed permanent magnet motor prototype. An analytical solution formula of rotor stresses in isotropic material was deducted. They also calculated the critical speed of the high-speed permanent magnet rotor system. The model structure diagram of their research institute is as shown in Fig. 9.6.

Professor Zhang Fengge et al. from the same research group were also contributing. With the same prototype, using analytic and finite-meta methods, they analyzed and compared the rotor strength under metal sheath and carbon fiber sheath. The results showed that carbon fiber sheath had higher rotor strength. Additionally, under the same speed, carbon fiber sheath only needed half of the thickness of metal sheath to achieve the same rotor strength. Their analysis also suggested that the strength of carbon fiber sheath was very little affected by temperature, while it was the contrast to metal ones. The motor could only be safe and reliable when the metal sheath can be designed to ensure the safety under cold and hot states.

Professor Zhu Changsheng et al. from Zhejiang University investigated on the thermal rotor strength carbon fiber sheath high-speed permanent magnet based on a 75 kW, 12,000 rpm fly-wheel energy storage high-speed permanent magnet synchronous motor. The structure of the rotor involved was a block-type permanent magnet structure. Setting out from Elasticity theory, they deducted the analytical expression of the rotor strength of the block-type permanent magnet. The analytical algorithm was used to calculate radial stress, tangential stress and equivalent stress in carbon fiber sheath, permanent magnet and shaft. These results were compared to those of finite meta-method. The results showed that the analytical solution can accurately calculate the stress distribution of the surface-mounted block-type permanent magnet rotor under the condition of stationary and high-speed rotation. There was no much deviation between the analytical solution and the finite-meta methods.

Cheng Wenjie et al. of Xi'an Jiaotong University derived the analytical formula of stress field, strain field and displacement field of interference fit of two-layer and three-layer rotors. They also verified the accuracy of the analytic formula with finite meta-methods. The design criteria of shrink-fit permanent magnet rotor strength in high-speed conditions were summarized, which provided a theoretical basis for the optimization. Taking a high-speed permanent magnet synchronous motor rated at 120 krpm and 10 kW as an example, two intensity design methods for the two commonly used interference-fit high-speed rotors were suggested.

9.3.3 Loss and Temperature Rise

High-speed permanent magnet synchronous motor has the advantage of high-power density due to high speed, but its loss also increases accordingly. The loss of high-speed motor can be categorized into stator loss and rotor loss. Whereas stator loss includes iron loss and copper loss, while rotor loss contains eddy current loss, iron loss and wind friction loss.

(1) Stator loss

The copper loss of the motor is relatively easy to calculate, the calculation, while not the iron consumption. This is because the frequency of the high-speed motor is relatively high. It can be several times or even ten times the work frequency. The loss coefficient of the stator material of the motor at high frequency can be significantly different, so the loss curve of the traditional silicon steel sheet of the motor is no longer applicable at high frequency. Furthermore, it is also hard to obtain the accurate loss curve at high frequency. In order to accurately calculate the iron consumption of high-speed motors, it is also necessary to precisely calculate the electromagnetic fields in high-speed motors. To reduce the iron consumption of the stator, Huang et al. designed a high-speed motor of 2 kW/20,000 rpm using a new soft magnetic composite (SMC) material, SOMALOYTM500, as the stator iron core material. With the help of 2-D and 3-D finite-meta calculation, they compared and analyzed the high frequency loss of new SMC materials and high-performance silicon steel sheets. They found out that when the operating frequency of the motor exceeded 1500 Hz, the iron core loss of the high-speed motor could be effectively reduced by using SMC material. Paulides et al. conducted the study based on a high-speed, high-power permanent magnet generator of 1.5 MW/20,000 rpm, which owned a fractional slot structure of 8 poles and 9 slots with an electrical frequency up to 1333 Hz. They compared the iron loss of two iron core lamination: One is with 3% of Si content and a thickness of 0.35, 0.2, 0. 127 mm. The other is of 6.5% of Si content and a thickness of 0.1 mm. The r2esults showed that for the motor with lower air gap magnetic density, the 6.5% SiFe (Silicon Iron) rather than the 3% SiFe could greatly reduce iron loss, and basically impose no impact on the electromagnetic performance of the motor.

(2) Rotor loss

With the accelerating rotors of high-speed permanent magnet motor, there will be sharply rising consumption and heat. Overheating will lead to an irreversible demagnetization, which will also affect the overall performance of the motor. Because of the poor thermal conditions of the rotor, cutting off the heat source is an effective measure to reduce the temperature rise. This makes it very important to study on how to deduct the rotor loss of high-speed permanent magnet motor. Since there is no slip of the rotors, the time harmonic of the current and the space harmonic caused by the cogging slot are shielded by the rotor sheath and permanent magnet. The rotor iron consumption can be so small that even possibly be neglected. The rotor loss is mainly concentrated in the wind on the surface, and the eddy current loss of permanent magnets and sheaths.

When rotating at high speed, the linear velocity on the surface of the rotor is very high, which brings great wind friction loss. This accounts for a large proportion of the whole rotor loss. In view of wind friction loss, Sarri and Jokinen [54] analyzed and experimented on the friction loss of high-speed motor rotors in detail. Wang et al. gave an approximate analytical calculation formula for high-speed motor's wind friction loss. Huynh et al. investigated on high-speed permanent magnet motor's wind friction loss and gave out their analytical calculation formula. They also designed a finite meta-calculation and experimental test of the loss of a 120 kW/70,000 rpm permanent magnet alternator.

Katsumi et al. used 3D finite meta-method to analyze the loss of built-in permanent magnet motor. They reduced the eddy current loss of the rotor by permanent magnet blocking. Since the carrier harmonics produced by the inverter are the main causes of the eddy current loss, they suggested that when the axial length of the permanent magnet was less than twice the depth of harmonic current transmission, it could effectively reduce the eddy current loss of high-speed motor rotor. Through the test and finite meta-analysis of the eddy current loss of the Nd-Fe-B permanent magnet of the permanent magnetic synchronous motor, the influences of different insulator parameters such as thickness and conductivity of the insulator on the current loss of rotor harmonics were analyzed. Masatsugu et al. studied on the eddy current loss of the surface-mounted high-speed permanent magnet synchronous motor and proved through finite meta-analysis that different pole-slot matching has a greater impact on the eddy current loss of rotors. They also drew the curve of the change of rotor eddy current with current harmonics and spatial harmonics. Nicola et al. figured out the relation between the eddy current loss and the pole-slot matching based on their studies on the influences of pole-slot matching of a high-speed permanent magnet synchronous motor on the distribution of the airgap magnetomotive force harmonics. After these trials, they came up with the best matching of the poles and the slots.

In view of the sheath loss of the surface-mounted permanent magnetic motor, with the help of analytical algorithms, Iran scholar M. Etemadrezaei calculated the eddy current loss of the carbon fiber sheath of the permanent magnet motor, and studied on the effect of placing copper shield between permanent magnet and sheath on eddy current loss of the motor. On this basis, the relation between copper shield thickness and motor loss were drew out. Professor Li Weili et al. of Harbin University of Science and Technology have researched on the effect of sheath conductivity on rotor loss. They proposed a sheath structure of copper coating, which was conducive to lessen the eddy current loss and temperature rise, meanwhile shielding the harmonics into permanent magnets. They also put forward that when the copper thickness remained the same, copper coating had a better effect than adding a copper shield outside the permanent magnet.

Bibliography

1. S. Zizhou, Z. Bo et al., Research of electromagnetic catapult system for UAV. Electr. Technol. Suppl. 57–58 (2010)
2. J.R. Kitzmiller, J.A. Pappas, S.B. Pratap, M.D. Driga, Single and multiphase compulsator system architectures: a practical comparison. IEEE Trans. Magn. **37**(1), 367–369 (2001)
3. Z. Jianyi, He. Wen, Review of research actuality and development directions of pulsed power technology. J. Mech. Electr. Eng. **25**(4), 1–3 (2008)
4. L. Dongli, Z. Jin et al., The applications of pulsed power in environmental engineering. High Volt. Eng. **28**(2), 35–37 (2002)
5. A. Shuping, L. Qi et al., Applications of high power pulse technology to increase production in low permeability reservoir. Inner Mongolia Petrochem. Ind. **16**, 3–5 (2009)
6. G. Li, Study on pulse corona non-thermal plasma reactor for SO_2 removal in flue gas. PhD dissertation, Dalian University of Technology, 2000
7. M. Akiyama, K. Minami, M. Watanabe, De-NO_x by bidirectional pulse corona discharge, in *IEEE Conference Record—Abstracts. The 27th IEEE International Conference on Plasma Science*, 2000
8. L. Jun, L. Xiucheng et al., Circuit topology of a new inductive storage pulsed-power supply to drive railgun. Power Syst. Technol. **33**(13), 80–81 (2009)
9. F.C. Beach, I.R. McNab, Present and future naval applications for pulsed power, in *IEEE Pulsed Power Conference*, 2005
10. W. Xinxin, Z. Zhuo et al., The charging circuit of repetitive MARX generator. High Volt. Eng. **23**(1), 37–39 (1997)
11. W. Ying, Strong pulsed power supply with inductive energy storage and its circuit breaker. Adv. Technol. Electr. Eng. Energy (1986)
12. L. Qing'ao, G. Min et al., Theory and technology of explosive magnetic flux compression generator with a helical winding. High Volt. Eng. **31**(6), 42–43 (2005)
13. H. Yan, Z. Guixin et al., *Fundamentals of Pulsed Power Technology* (Tsinghua University Press, 2010)
14. W. Ying et al., *Pulsed Power Science and Technology* (Beihang University Press, 2010)
15. W. Ying, R.A. Marshall, C. Shukang, *Physics of Electric Launch* (Science Press, Beijing, 2004).
16. G. Zhang, K. Yu, Research and design of air-core compensated pulsed alternator. Master Thesis of Mechanical and Electrical Engineering, Huazhong University of Science and Technology, 2008
17. X. Shangang, Z. Shichang, S. Liming, Compensated pulsed alternator. Adv. Technol. Electr. Eng. Energy **1**, 16–22 (1990)
18. W.F. Weldon, Compensated pulsed alternator, U.S. Patent 4200831, 29 April 1980
19. J.R. Kitzmiller, S.B. Pratap, M.D. Driga, An application guide for compulsators. IEEE Trans. Magn. **31**(1), 285–287 (2003)

S. Wu and S. Cui, *Pulsed Alternators Technologies and Application*,
https://doi.org/10.1007/978-981-33-4224-8

20. W. Shaopeng, Research on design methods and key technologies of air-core compensated pulsed alternator. Ph.D. dissertation in Electrical Engineering, Harbin Institute of Technology, 2011
21. L. Jun, Y. Ping et al., Electromagnetic gun technology and its development. High Volt. Eng. **40**(4), 1052–1064 (2014)
22. L. Ge, *Design and Experiments of the Passive Compulsator* (Institute of Plasma Physics, Chinese Academy of Sciences, Hefei, 1993), pp. 1–5
23. L. Ge, Analytic calculation of the air-gap armature coil inductance of the passively compensated pulsed alternator. Nucl. Fusion Plasma Phys. **20**(2), 125–128 (2000)
24. L. Ge, W. Yong, L. Baohua et al., Experimental study of 25 MW rotary flux compression pulsed alternator. J. Univ. Sci. Technol. China **30**(5), 561–566 (2000)
25. Z. Ke, *Design and Experiments of the 25 MW Passive Compulsator* (Huazhong University of Science and Technology, 2004)
26. M. Brennan, W.L. Bird, J.H. Gully et al., The mechanical design of a compensated pulsed alternator prototype, in *2nd IEEE International Pulsed Power Conference*, Lubbock, TX, USA, 1979, pp. 392–397
27. S.B. Pratap, W.L. Bird, G.L. Godwin et al., A compulsator driven rapid-fire EM gun. IEEE Trans. Magn. **20**(2), 211–214 (1984)
28. W.A. Walls, M.L. Spann, S.B. Pratap et al., Design of a self-excited, air-core compulsator for a skid-mounted repetitive fire 9 MJ railgun system. IEEE Trans. Magn. **25**(1), 574–579 (1989)
29. J.R. Kitzmiller, S.B. Pratap, T.A. Aanstoos et al., Optimization and critical design issues of the air-core compulsator for the cannon caliber electromagnetic launcher system (CCEML). IEEE Trans. Magn. **31**(1), 61–66 (1995)
30. J.R. Kitzmiller, K.G. Cook, J.J. Hahne et al., Predicted vs. actual performance of a model scale compulsator system. IEEE Trans. Magn. **37**(1), 362–366 (2001)
31. D. Putley, Analysis and modeling of the culham experimental compulsator, in *6th IEEE International Pulsed Power Conference*, Arlington, VA, USA, 1987, pp. 514–517
32. C.R. Spikings, D. Putley, Simulation of selective passive compensation. IEEE Trans. Magn. **27**(1), 415–420 (1991)
33. P.V. Vassioukevitch, Iron-core compulsator, in *9th IEEE International Pulsed Power Conference*, Albuquerque, NM, USA, 1993, pp. 224–227
34. H. Akiyama, Pulsed power in Japan, in *10th IEEE International Pulsed Power Conference*, Albuquerque, NM, USA, 1995, pp. 13–16
35. J.F. Eastham, M.J. Balchin, R.J. Hill-Cottingham et al., Numerical analysis techniques applied to a model iron-cored compulsator. IEEE Trans. Magn. **31**(1), 587–592 (1995)
36. E.A. Lomonova, S.R. Miziurin, The theoretical investigations and mathematical models of the compulsators, in *7th International Conference on Electrical Machines and Drives*, Durham, UK, 1995, pp. 131–135
37. J. Jaewon, L. Younghyun, Y. Kyungseung et al., Overview of ETC program in Korea. IEEE Trans. Magn. **37**(1), 39–41 (2001)
38. Curtiss-Wright, Pulsed Power Systems/EM Gun [OL]. https://apsd.cwfc.com/defensegov/spokes/02_pulsedpower-em.htm
39. B. Joe, *Center for Electromechanics Energy Storage and Pulsed Power Research*
40. W. Shaopeng, *Research on Design Methods and Key Technologies of Air-Core Compensated Pulsed Alternator* (Harbin Institute of Technology, 2011)
41. Z. Weiduo, *Key Problems of Air-Core Compensated Pulsed Alternators with Highly Transferrable Energy Density* (Harbin Institute of Technology, 2015)
42. A. Hughes, T.J.E. Miller, Analysis of fields and inductances in air-cored and iron-cored synchronous machines. Proc. Inst. Electr. Eng. **124**(2), 121–126 (1977)
43. S.J. Oh, M.D. Driga, Fast discharging inertial energy storage systems for industry applications: electromagnetic considerations, in *IEEE Pulsed Power Conference*, Monterey, CA, USA, 1999, pp. 1283–1286

44. S.B. Pratap, M.D. Driga, Compensation in pulsed alternators. IEEE Trans. Magn. **35**(1), 372–377 (1999)
45. W.A. Walls, *Advanced Compulsator Topologies and Technologies* (The University of Texas at Austin, 2002)
46. C. Ye, K. Yu, Z. Lou, Y. Pan, Investigation of self-excitation and discharge processes in an air-core pulsed alternator. IEEE Trans. Magn. **46**(1), 150–154 (2010)
47. J.A. Mallick, M. Crawford, Determining pulsed alternator thyristor converter firing angles to produce a desired launcher current. IEEE Trans. Magn. **41**(1), 322–325 (2005)
48. S.K. Goswami, Synchronous-machine sudden 3-phase short-circuit. Analysis by Norton's, constant-flux-linkage and Thévenin's theorems. Proc. Inst. Electr. Eng. **118**(10), 1459–1466 (1971)
49. L. Kefu, Fully compensated pulsed alternator and its system—pulsed power supply of high-power miniaturized electric guns. Ph.D., Huazhong University of Science and Technology, Wuhan, 1999, pp. 46–47
50. Y. Kexun, Y. Caiyong, P. Yuan, Study on the excitation condition of compensated pulsed alternator, in *International Conference on Electrical Machines and Systems 2008*, Wuhan, China, 2008, pp. 3544–3548
51. W. Yongtian, M. Dawei, *Heat Exchange in the Motor* (China Machine Press, 1998)
52. C. Shikun, *Motor Design* (China Machine Press, 2000), pp. 133–134
53. W. Fujun, *Computational Fluid Dynamics Analysis—CFD Software Principles and Applications* (Tsinghua University Press, 2004), pp. 1–23
54. A.I. Borisenko, *Aerodynamics and Heat Transfer in Motors* (China Machine Press, 1985), pp. 114–124
55. K.S. Ball, B. Farouk, V.C. Dixit, Experimental study of heat transfer in a vertical annulus with a rotating inner cylinder. Int. J. Heat Mass Transf. **32**(8), 1517–1527 (1989)
56. V. Hatziathanassion, J. Xypteras, G. Archontoulakis, Electrical-thermal coupled calculation of an asynchronous machine. Arch. Elektrotech. Berlin **77**(2), 117–122 (1994)
57. C. Junci, L. Weili, C. Shukang, Temperature field calculation and associated factor analysis of induction motor with compound cage rotor. Proc. Chin. Soc. Electr. Eng. **30**, 96–103 (2008)
58. Y. Fei, *Temperature Calculation and Cooling System Design of Permanent-Magnet Electric Motors* (Shenyang University of Technology, 2007)
59. H. Guozhi, F. Fengli, *Technical Manual for Y2 Three-phase Asynchronous Motor* (China Machine Press, 2005), pp. 138–152
60. Q. Weichang, *Theoretical Basis for Strength Calculation in Motor Design* (Science and Technology Press, Anhui, 1992), pp. 1–4
61. Y. Song, *Mechanical Analysis of the Rotor about Compensated Pulsed Alternator* (Huazhong University of Science and Technology, Wuhan, 2004), pp. 17–41
62. S.K. Ha, H.M. Jeong, Y.S. Cho, Optimum design of thick-walled composite rings for an energy storage system. J. Compos. Mater. **3**(29), 851–873 (1998)
63. G. Nengping, X. Yuanming, M. Tianxiang, Design of composite flywheel structure and its stress analysis. Compos. Mater. **19**(1), 113–116 (2002)
64. M. Ozdemir, Electrodynamic forces in compensated pulsed electrical machines. Doctor dissertation, The University of Texas at Austin, Austin, 2002, pp. 136–141
65. J.W. Kim, J.H. Lee, H.G. Kim, Reduction of residual stresses in thick-walled composite cylinders by smart cure cycle with cooling and reheating. Compos. Struct. **75**(1–4), 261–266 (2006)
66. S.K. Ha, H.T. Kim, Effects of rotor sizes and epoxy system on the process-induced residual strains within multi-ring composite rotors. J. Compos. Mater. **38**(10), 871–885 (2004)
67. S.K. Ha, M.H. Kim, Design and spin test of a hybrid composite flywheel rotor with a split type hub. J. Compos. Mater. **40**(23), 2113–2130 (2006)
68. G. Genta, *Kinetic Energy Storage: Theory and Practice of Advanced Flywheel Systems* (Cambridge University Press, London, UK, 1985), pp. 128–147
69. *Composite Structure Design* (Chemical Industry Press, Beijing, 2001), pp. 44–52

70. S. Guanlin, H. Gengkai, *Composite Mechanics* (Tsinghua University Press, Beijing, 2006), pp. 56–62
71. G. Portnov, A.N. Uthe, I. Cruz et al., Design of steel-composite multirim cylindrical flywheels manufactured by winding with high tension and in situ curing. Part. 1. Basic relations. Mech. Compos. Mater. **41**(2), 139–152 (2005)
72. G. Portnov, A.N. Uthe, I. Cruz et al., Design of steel-composite multirim cylindrical flywheels manufactured by winding with high tensioning and in situ curing. Part. 2. Numerical analysis. Mech. Compos. Mater. **41**(3), 241–254 (2005)
73. K.H. Sung, J.H. Kim, Y.H. Han, Design of a hybrid composite flywheel multi-rim rotor system using geometric scaling factors. J. Compos. Mater. **42**(8), 771–785 (2008)
74. S.K. Ha, H.H. Han, Y.H. Han, Design and manufacture of a composite flywheel press-fit multi-rim rotor. J. Reinf. Plast. Compos. **27**(9), 953–965 (2008)
75. X. Li, *Research on the Control Technology of Modular Inertial Energy Storage Pulsed Power Supply* (Harbin Institute of Technology, 2018)
76. W. Ying, X. Feng, *The Principle of Electric Gun* (National Defense Industry Press, Beijing, 1995).
77. Z. Changjun, S. Zizhou, Z. Tao et al., Design and simulation of a extra large muzzle kinetic energy railgun. J. Gun Launch Control **3**, 10–14 (2013)
78. Xiangyang, G. Gang, Z. Jiange, The overview of the electrothermal-chemical launch technology abroad. Ship Sci. Technol. **29**(A01), 159–162 (2007)
79. G. Wei, Z. Tao, F. Wei, Simulation and analysis of magnetic coupling of synchronous induction coilgun. J. Gun Launch Control **35**(1), 10–14 (2014)
80. D. Putley, Current pulse requirements for rail launchers, in *8th IEEE International Pulsed Power Conference*, San Diego, CA, USA, 1991, pp. 767–770
81. S.B. Pratap, M.D. Driga, W.F. Weldon et al., Future trends for compulsators driving railguns. IEEE Trans. Magn. **22**(6), 1681–1683 (1986)
82. Y. Yu, L. Liang, *Pulsed Power Switch Components and Its Application* (China Machine Press, 2010)
83. L. Bo, *Study on Thyristor Application in Pulse Power System* (Huazhong University of Science and Technology, 2009)
84. X. Yongxiang, Research on control and rotor loss of a high-speed PM synchronous motor with one discrete hall sensor. Ph.D. thesis from Harbin Institute of Technology, 2005, p. 2
85. Z. Fengso, Investigation on eddy-current loss in high speed PM BLDC motors. Ph.D. dissertation from Zhejiang University, 2008, pp. 5–10
86. Z. Yonggao, Novel high-speed 8000 rpm DC motor. Electr. Mach. Technol. **04**, 6–29 (1987)
87. N. Morita, T. Ueno, M. Mori et al., Comparison of brush contact drop characteristics with copper ring and steel ring, considering spark and collector ring hazard, in *53rd IEEE Holm Conference on Electrical Contacts*, Pittsburgh, PA, USA (IEEE, 2007), pp. 300–305
88. N. Morita, T. Ueno, T. Otaka et al., Comparison of brush dynamic operation characteristics for turbine generator steel collector ring, in *53rd IEEE Holm Conference on Electrical Contacts*, Pittsburgh, PA, USA (IEEE, 2007), pp. 205–210
89. W.R. Snow, R.L. Willig, Design criteria for brush commutation in high speed traveling wave coilguns. IEEE Trans. Magn. **27**(1), 654–659 (1991)
90. H.G. Rylander, J.H. Gully, Z. Eliezer, Limiting design parameters for future high speed, high current density sliding electrical contacts, in *Proceedings of the Thirty Fourth Meeting of the IEEE Holm Conference on Electrical Contacts* (IEEE, 1988), p. 247
91. C.P. Chen, R.A. Burton, Thermoelastic effects in brushes with high current and high sliding speeds. Wear **59**(1), 277–288 (1979)
92. J. Pyrhonen, J. Nerg, P. Kurronen et al., High-speed high-output solid-rotor induction-motor technology for gas compression. IEEE Trans. Ind. Electron. **57**(1), 272–280 (2010)
93. M. Centner, R. Hanitsch, U. Schafer, Comparison of high-speed induction motors employing cobalt-iron and silicon electrical steel, in *18th International Conference on Electrical Machines*, Vilamoura, Algarve, Portugal (IEEE, 2008), pp. 1–6

94. W. Jiqiang, Study on the mechanical and electromagnetic properties of high-speed permanent magnet machine. Ph.D. thesis from Shenyang University of Technology, 2007, p. 4
95. M. Ahrens, U. Bikle, R. Gottkehaskamp et al., Electrical design of high-speed induction motors of up to 15 MW and 20000 rpm, in *2002 International Conference on Power Electronics, Machines and Drives*, Bath, UK (Institution of Engineering and Technology, 2002), pp. 381–386
96. M.A. Rahman, A. Chiba, T. Fukao, Super high speed electrical machines—summary, in *IEEE Power Engineering Society General Meeting*, Denver, CO, USA (Institute of Electrical and Electronics Engineers Inc., 2004), pp. 1272–1275
97. Q. Li, Research and practice on integrated starter alternator damper system for vehicles based on switched reluctance motor. Ph.D. thesis from Nanjing University of Aeronautics and Astronautics, 2007, pp. 2–10
98. S. Jie, Sensorless position control and neutral network control of SRM. Ph.D. thesis from Tianjin University, 2007, pp. 7–9
99. L. Jianwu, Research of the switched reluctance drive system for in-wheel driving electric vehicle. Ph.D. thesis from Huazhong University of Science and Technology, 2007, pp. 2–5
100. H. Ku, T. Wiehert, B. Szymanski, Design of a high speed switched reluctance motor for spindle drive, in *5th International Conference-Workshop Compatibility in Power Electronics*, Gdansk, Poland (IEEE, 2007), pp. 1–5
101. L. Morel, H. Fayard, R.V. Fos et al., Study of ultra high speed switched reluctance motor drive, in *35th IAS Annual Meeting and World Conference on Industrial Applications of Electrical Energy*, Rome, Italy (IEEE, 2000), pp. 87–92
102. Z. Xuyong, Comparison and analysis of high-speed switched reluctance motor with different structure. Electr. Mach. Technol. **1**, 37–38 (2008)
103. F. Dong, Y. Yan, D. Zhiquan et al., The key technology on designing a high-speed bearingless switched reluctance motor. Electr. Mach. Control **10**(06), 547–552 (2006)
104. Q. Xiangwen, S. Jian, L. Cheng, Research on digital controller of high-speed switched reluctance motor. Mech. Electr. Technol. **03**, 43–46 (2008)
105. C. Haijin, Design and verification of digital controller used for high speed switched reluctance motor. Micromotors **40**(8), 49–53 (2007)
106. Z. De'an, Research on advanced control method of switched reluctance motor in ISAD system of vehicles. Ph.D. thesis from Nanjing University of Aeronautics and Astronautics, 2007, pp. 7–8
107. X. Mei, Research on nonlinear modeling and intelligent control strategy of switched reluctance motors. Ph.D. thesis from Tianjin University, 2008, pp. 5–13
108. Y. Xiuke, Field-circuit-motion coupled analysis for switched reluctance motor. Ph.D. thesis from Shenyang University of Technology, 2005, pp. 11–13
109. T.-H. Liu, C.-K. Hung, D.-F.A. Chen, A matrix converter-fed sensorless PMSM drive system. Electr. Power Compon. Syst. **33**(8), 877–893 (2005)
110. Z. Meng, C. Sun, Y. An et al., Non-interacting control of PMSM based on exact linearization via state variable feedback, in *2009 International Workshop on Intelligent Systems and Applications*, Wuhan, China (IEEE, 2009), pp. 1–4
111. C. Bian, S. Ren, S. Yan et al., Optimal V/F control of super high-speed PMSM and its application, in *Sixth International Symposium on Instrumentation and Control Technology: Sensors, Automatic Measurement, Control, and Computer Simulation*, Beijing, China (SPIE, 2006), pp. 1–6
112. I. Takahashi, T. Koganezawa, G. Su et al., A super high speed PM motor drive system by a quasi-current source inverter. IEEE Trans. Ind. Appl. **30**(3), 683–690 (1994)
113. A.S. Nagorny, N.V. Dravid, R.H. Jansen et al., Design aspects of a high speed permanent magnet synchronous motor/generator for flywheel applications, in *IEEE International Conference on Electric Machines and Drives*, San Antonio, TX, USA (IEEE, 2005), pp. 635–641
114. T. Shelley, Direct coupled turbines power the future. Eureka **15**(9), 35–36 (1995)

115. C. Zwyssig, M. Duerr, D. Hassler et al., An ultra-high-speed, 500000 rpm, 1 KW electrical drive system, in *4th Power Conversion Conference*, Nagoya, Japan (IEEE, 2007), pp. 1577–1583
116. H. Yunkai, Y. Li, H. Qiansheng, Design considerations for high-speed permanent magnet motors. Micromotors **39**(08), 6–9 (2006)
117. Y. Li, H. Qiansheng, C. Yang et al., Analysis and calculation of the iron losses of high speed permanent motor. Small Spec. Electr. Mach. **03**, 11–13 (2008)
118. Y. Li, L. Hexiang, H. Qiansheng, Optimization design of high-speed permanent BLDC motor. Mach. Build. Autom. **06**, 117–119 (2009)
119. W. Jiqiang, W. Fengxiang, K. Xiaoguang, Design and analysis of electromagnetic properties for high speed PM generator. Proc. CSEE **28**(20), 105–110 (2008)
120. Z. Fengzheng, S. Jianxin, W. Kai et al., Research on topology of high-speed PM brushless DC motors. Small Spec. Electr. Mach. **10**, 1–6 (2007)
121. Z. Xiaochen, *Temperature Distribution and Structure Optimization of High-Speed PM Generator* (Harbin Institute of Technology, Harbin, 2012), pp. 30–52
122. Z. Xiaochen, L. Weili, Q. Hongbo, C. Shukang, Calculation of electromagnetic field and temperature field in super high speed permanent magnet generator with composite structures. Proc. CSEE **31**(30), 85–92 (2011)
123. V. Gradinaru, L. Tutelea, I. Boldea, 25 kW, 15 krpm, 6/4 PMSM: optimal design and torque pulsation reduction via FEM, in *Proceedings of Optimization of Electrical and Electronic Equipment, 2008. OPTIM'2008*, Brasov, 2008, pp. 249–256
124. S.-M. Jang, H.-W. Cho, S.-H. Lee et al., Analysis of cogging torque for high speed motor/generator, in *Proceedings of Electrical Machines and Systems, 2003. ICEMS'2003*, Beijing, China, 2003, vol. 1, pp. 205–207
125. S. Jang, J. Choi, H. Cho et al., Electromagnetic analysis of high speed machines with diametrically magnetized rotor for flywheel battery applications. Intermag (2006)
126. J.M. Yon, P.H. Mellor, R. Wrobel et al., A semi-permeable containment sleeve for high-speed PM machines, in *Proceedings of Power Electronics, Machines and Drives* (2010), pp. 1–6
127. A. Binder, T. Schneider, M. Klohr, Fixation of buried and surface mounted magnets in high-speed permanent magnet synchronous motors, in *Proceedings of Industry Applications Conference*, 2005, pp. 2843–2848
128. A. Borisavljevic, H. Polinder, J.A. Ferreira, Enclosed sure design for a high-speed permanent magnet rotor, in *Power Electronics, Machines and Drives*, 2010, pp. 1–6
129. Z. Fengge, D. Guangguang, W. Tianxuan, H. Na, Rotor strength analysis of high-speed permanent magnet under different protection measures, in *Proceedings of the CSEE (Supplement)*, vol. 33, pp. 195–202
130. W. Baojun, B. Liuxin, C. Liangliang, Y. Pingxi, Z. Changsheng, Strength analysis of a surface mounted high speed permanent magnetic machine rotor with carbon fiber bandage. J. Zhejiang Univ. (Eng. Sci.) **47**(12), 2101–2110 (2013)
131. W. Baojun, Design and analysis of permanent magnet motors for high-speed flywheel energy storage system. Master's thesis from Zhejiang University, Hangzhou, 2013
132. C. Wenjie, G. Haipeng, F. Sheng et al., Rotor strength analysis of high-speed permanent magnet synchronous motors. Proc. CSEE **32**(27), 87–94 (2013)
133. Y.K. Huang, Q.S. Hu, J.F. Zhao et al., Comparative study of high-speed PM motors with laminated steel and soft magnetic composite cores, in *Proceedings of Industry Applications Conference, 2007*, New Orleans, 2007, pp. 67–72
134. J.J.H. Paulides, G.W. Jewell et al., An evaluation of alternative stator lamination materials for a high-speed, 1.5 MW, permanent magnet generator. IEEE Trans. Magn. **40**(4), 2041–2043 (2004)
135. J. Saari, *Thermal Analysis of High-Speed Induction Machines* (Helsinki University of Technology, Juha Saari, 1998), pp. 1–73
136. T. Jokinen, J. Sarri, Modeling of the coolant flow with heat flow controlled temperature sources in thermal networks. IEEE Proc. Electr. Power Appl. **144**(5), 338–342 (1997)

137. W. Jiqiang, W. Fengxiang, Y. Tao, Analysis of rotor losses for a high speed PM generator, in *Proceedings of the Eighth International Conference on Electrical Machines and Systems*, 2005, pp. 889–892
138. C. Huynh, L.P. Zheng, D. Acharya, Losses in high speed permanent magnet machines used in microturbine applications. J. Eng. Gas Turb. Power **131**(02), 1–6 (2009)
139. Y. Katsumi, A. Atsushi, Loss investigation of interior permanent magnet motors considering carrier harmonics and magnet eddy currents. IEEE Trans. Ind. Appl. **45**(2), 659–665 (2009)
140. N. Masatsugu, K. Haruyuki, K. Mitsuhiro, A study on eddy current losses in rotors of surface permanent magnet synchronous machines. IEEE Trans. Magn. **5**(04), 1696–1702 (2004)
141. B. Nicola, B. Silverio, F. Emanuele, A general approach to determine the rotor losses on three phase fractional slot PM machines. IEEE Trans. Magn. **05**(07), 634–641 (2007)
142. M. Etemadrezaei, J. Wolmarans, H. Polinder, J.A. Ferreira, Precise calculation and optimization of rotor eddy current losses in high speed permanent magnet machine, in *International Conference on Electrical Machines*, 2012, pp. 1399–1404
143. Q. Hongbo, Research on eddy current loss optimization and temperature distribution in high speed permanent magnet generator. Ph.D. thesis, Harbin University of Science and Technology, Harbin, 2014

The manufacturer's authorised representative in the EU is Springer Nature Customer Service Centre GmbH, Europaplatz 3, 69115 Heidelberg, Germany. If you have any concerns regarding our products, please contact ProductSafety@springernature.com

Printed and bound by CPI Group (UK) Ltd, Croydon, CR0 4YY

30/07/2026

02181994-0003